Ageless

Ageless

The New Science of Getting Older
Without Getting Old

ANDREW STEELE

RANDOM HOUSE LARGE PRINT

All rights reserved. Published in the United States of America by Random House Large Print in association with Doubleday, a division of Penguin Random House LLC, New York, and distributed in Canada by Penguin Random House Canada Limited, Toronto.

Originally published in hardcover in Great Britain by Bloomsbury Publishing, Bloomsbury Publishing Plc, London, in 2020.

Cover photograph by cris180 / Getty Images
Cover design by Emily Mahon

The Library of Congress has established a Cataloging-in-Publication record for this title.

ISBN: 978-0-593-21479-4

www.penguinrandomhouse.com/large-print-format-books

FIRST LARGE PRINT EDITION

Printed in the United States of America

10 9 8 7 6 5 4 3 2 1

This Large Print edition published in accord with the standards of the N.A.V.H.

Contents

Part III LIVING LONGER

Ageless

Introduction

Wrinkled, toothless and ambling with a ponderous gait, it seems unlikely at first glance that the giant Galápagos tortoise could teach us anything about aging gracefully. They are eponymous inhabitants of the remote Galápagos Islands, a volcanic archipelago in the Pacific which draws its name from the old Spanish word **galápago,** meaning "tortoise." These cumbersome reptiles can weigh over 400 kilograms, and take decades to reach maturity on a diet of leaves and lichens.

The Galápagos Islands were made famous after Charles Darwin visited in 1835, and their unique flora and fauna inspired his theory of evolution by natural selection. The giant tortoises were one of the many unusual species which greeted him, and he collected several specimens to return to England for further study. One of those tortoises, Harriet, went on to become the oldest recorded Galápagos tortoise—she finally succumbed to a heart attack in 2006 at the ripe old age of 175, surviving Darwin by well over a century.

However, it's not their impressive longevity which is most interesting when it comes to the biology of aging; you could make the case that these tortoises' extreme long life is down to their slow pace of living rather than any special biological abilities—the candle which burns half as bright burns twice as long, so to speak. Far more interesting is that Galápagos tortoises, along with a few other species of tortoises and turtles, some fish, salamanders, and a handful of other, stranger creatures, display what's called "negligible senescence"—a negligible loss of capacity as they age. Negligibly senescent animals have no obvious impairments of movement or senses as they get older, and they experience no age-related decline in fertility. Harriet was likely pretty much as sprightly at 170 as she was at 30, at the height of Queen Victoria's reign—which is to say, not very; she was a giant tortoise, after all.

We humans are not so lucky. As we advance in years, we become wrinkled, frail and at increased risk of illness. Perhaps the most striking way to summarize our increasing frailty is to examine how our risk of death changes with time. Tortoises, being negligibly senescent, have a risk of death which is more or less constant with age: as adults, they have an approximately 1–2 percent chance of dying every year. We, by contrast, have a risk of death which doubles every eight years. This doesn't start out so bad: aged 30, your odds of dying that year are less than 1 in 1,000. However, if you keep on doubling something it can start small but, eventually, get very large very quickly: at 65, your risk of death that year is 1 percent; at 80, 5 percent; and by 90, if you make it that far, your odds of not making your 91st birthday are a sobering one in six. There is some evidence that this relationship flattens out after the age of 105 or so, meaning that these exceptionally long-lived people might have technically stopped aging—but, with odds of death around 50 percent per year by then, they might wish it had flattened out slightly sooner.

We enjoy a relatively long period of fitness, perhaps five or six decades where our risk of death, disease and disability is fairly low, before a precipitous rise in old age. Aging happens to all of us and growing old brings experience and wisdom; to do so gracefully is something to aspire to. Since the dawn of life, aging has been a natural part of being alive. Thus, the word "aging" comes with a variety of connotations, not all of them negative. But, from a biological perspective, perhaps the best (and certainly the simplest) definition of aging is the exponential increase in death and suffering with time.

By this biological definition, tortoises don't

age—they are, quite literally, ageless. This means that negligible senescence is sometimes known by another, more beguiling name: "biological immortality." How do tortoises get old without getting elderly? And, with the help of science, could we become ageless, too?

Modern science, particularly in the last couple of decades, has made huge strides in both our understanding of and our ability to intervene in the aging process. Aging impacts our biology at every level, from molecules, to cells, to organs, to whole systems. I want to show you what is happening biologically as we age, and how understanding its scientific implications could lead to a complete transformation of medical care.

Understanding aging could have enormous implications because it is by far the world's leading cause of death and suffering. While that might sound counterintuitive, looking at aging as a biological process makes the logic inescapable. As we age, our bodies accrue a familiar array of changes—from the superficial, like gray hair, wrinkles and elongating noses and ears, to the life-changing, like frailty, loss of memory and the risk of deadly diseases. The fundamental reason that our risk of death rises so swiftly is a rapid, synchronized increase in the odds of age-related disease. Even if you're relaxed about death itself—we all have to go sometime, after all—then this risk of death is still a proxy for years of suffering at the hands of disability and disease which we would probably all rather avoid.

Every extra year you live, your risk of getting cancer, heart disease, stroke, dementia and many more terrible conditions increases inexorably. Doctors and scientists call anything that increases your chance of getting a disease a "risk factor": these are things like smoking, being overweight, not getting enough exercise and so on. But simply getting older, no matter how well you live, dwarfs their effects. In fact, being old is the single biggest risk factor for **all** of the diseases just mentioned. An 80-year-old is 60 times more likely to die than a 30-year-old—so, too, are they 30 times more likely to get cancer, and 50 times more likely to get heart disease. Having high blood pressure doubles your risk of having a heart attack; being 80 rather than 40 multiplies your risk by ten. Dementia is extremely rare under the age of 60 but, after that, risk doubles every five years—even faster than the rising risk of death. From the perspective of disease risk at least, it's better to be an overweight, heavy-drinking, chain-smoking 30-year-old than a clean-living 80-year-old.

The end result of these synchronized increases in risk is a huge burden of disease. Half of people aged 65 have two or more long-term conditions. The average 80-year-old suffers from around five different diseases and takes a similar number of different types of medication for them. Even though it's entered into idiom, it's not really possible to "die of old age": instead, these diseases develop and progress until eventually one of them becomes severe enough to take your life.

Then, there are changes that mean it's easier for you to get ill in other ways, and make it dramatically worse when you do. For example, as we age, our immune system weakens, and loses its edge at fighting infection. That can mean that a bout of the flu, which meant a week in bed as a youth, might be the end of you as a senior citizen. Similarly, a broken bone might mean an annoying stint hobbling around wearing a plaster cast for a young person, but weeks in a hospital bed and a debilitating loss of muscle mass for an older person, making returning to normal life afterward difficult or impossible.

Finally, there are the symptoms which quietly eat away at quality of life: a loss of mental acuity, forgetfulness or increasing anxiety which doesn't meet the threshold for dementia; a decrease in muscle strength, plus conditions like rheumatism and arthritis, which reduce our ability to get about or do things around the house on our own; and embarrassing and inconvenient changes, from impotence to incontinence. Together, even if you've not got a specific disease to diagnose, these combine to sap our independence, self-esteem, enjoyment of life and contribution to society as our years advance.

We're used to considering every item in this laundry list of diseases and dysfunctions as an individual condition, largely separate from the others. Our approach to medicine is correspondingly idiosyncratic, with drugs and operations for cancer and heart disease, vaccines to prevent infections, and walking sticks and social care to help with everyday life.

The root cause, the aging process itself, we ignore entirely.

The universality of aging means that it has enormous consequences: imagine the life-changing effects of getting older on an individual—the decreasing independence and quality of life, coupled with the dramatically increasing risk of disease and death—multiplied by billions of people. Nor does it just affect those who are old and infirm today: most of us at some stage will need to care for an elderly friend or relative. The effects of aging reverberate throughout society, impacting all of our lives.

About 150,000 people die every day on Earth. Over 100,000 of them die because of aging. This means that, globally, aging is responsible for more than two-thirds of deaths—and over 90 percent in rich countries. Tens of millions suffer over years or decades as their health deteriorates. A natural disaster on this scale would be utterly unprecedented. A huge and immediate international relief effort would be mounted, even if success was uncertain. If a disease with these symptoms were to suddenly arise in a previously ageless civilization, Herculean efforts to cure it would begin as soon as possible.

But its ubiquity also means that aging is the

default; its inevitability makes it invisible. We see individual tragedies as friends and relatives age, and acknowledge the horror of the specific diseases that afflict them—but society is collectively casual about aging itself. This rolling worldwide pandemic of death and suffering goes unrecognized, too large to grasp, obscured by its own enormity.

We humans are beset by a cocktail of cognitive biases which emphasize the here and now and minimize the distant future. Most of us don't save enough for our retirement and find it hard to stick to diets or exercise regimens. Human beings are also wired for optimism. We might picture ourselves gray-haired, retired, taking up new hobbies or playing with our grandchildren. We don't picture being in the hospital with an IV line and a bladder catheter. Research shows that we don't deny the existence of cancer or heart attacks—just that few people believe it will happen to **them.** We also tend to extrapolate from previous experience. Thankfully, most of us don't experience multiple simultaneous chronic diseases before we get old; when we picture our retirement, we don't imagine being ill simply because we've not got much to go on.

We are also insulated from the consequences of aging when it happens to other people. The oldest and most unwell are tucked away in hospitals and nursing homes, hidden from view. When we are children, our grandparents are often kind, wrinkly souls whose health problems we aren't really aware of. Even as adults with fledgling careers and a young family, we rarely get involved caring for older friends and relatives. That responsibility usually falls either to our own parents stepping in to look after their mom and dad or to our grandparents themselves taking care of one another. All this means we usually don't get to see the full picture until our parents, or even our partners, need taking care of—by which time we're starting to age ourselves. While these are gross generalizations which will vary from family to family, they are borne out by statistics: a U.S. survey found that those caring for someone over 65 had an average age of 63 themselves. We can easily make it through the first four, five or even six decades of life without having to confront what aging means— which makes it all the easier to put from our minds.

If we do think about what our lives might be like in 10, 20, 50 years' time, we can allay our anxieties by telling ourselves that we are fortunate. Aging is a curse that we in the rich world are perversely lucky to have; we live long enough for it to be a problem. Better to live a long life and die from a heart attack than die in childhood of malaria, right? It is, of course—and the fact that deaths from diseases like malaria are in significant part preventable makes their continued death toll a moral indictment— but the bittersweet news is that age-related diseases outgun other causes of death in over three-quarters of countries around the world.

Global life expectancy was 72.6 years in 2019, and

it's rising. If you already knew that, you're in the minority: in spite of our optimism about our own individual futures, surveys show that most people are pessimistic about the state of the world, and assume that life expectancy is 10 or even 20 years lower. Most of us imagine a large "developing world" where birth rates and death rates are high—this is, after all, what we were taught in school. The reality is that most countries are approaching the developed world, in life expectancy if not in wealth. This is astonishing progress, and well worth celebrating— we've beaten many deadly infectious diseases into submission, and improved quality and quantity of life worldwide. The flip side is that 70 is easily old enough to feel the effects of aging—this is another way to understand why age-related conditions are the greatest cause of death and suffering worldwide.

Aging is also a crisis which is snowballing as development continues and our global population gets older: even if it somehow fails to meet the definition of a global challenge now, it surely will do so in decades to come. The question is, what can we do?

The answer, thankfully, is biology. It all started in the 1930s, with a breakthrough that changed scientific history. There was a burgeoning interest in the new field of nutrition, and researchers were starting to wonder about the effects of food on growth and lifespan. Scientists took three groups of rats, one allowed to eat what they liked, and the other two on significantly more frugal diets, all while meticulously ensuring that they got all the nutrients they needed. The rats eating less were smaller than those in group one but, as the experiment progressed, it became obvious that their size wasn't the only thing being affected by their reduced rations. One by one, the rats eating what they liked got old and died, while the dieting rats kept going, and going. And these hungry rats didn't hobble along in ill health, gray-haired and cancer-ridden, somehow unable to muster the energy even to die after their better-fed counterparts had done so. The calorie-restricted animals were healthier and less frail for longer, too. It seemed almost as though eating less slowed down the aging process itself.

It turns out that this wasn't a fluke or experimental error. We've since tried dietary restriction in creatures from all over the tree of life, with astonishing generalizability: single-celled yeast (the fungus used in baking and brewing beer), worms, flies, fish, mice, dogs and many more all live longer and healthier if fed significantly less than normal. They are more active, and suffer less from the ailments of aging, from cancer to heart trouble (at least in those creatures that have hearts). Dietarily restricted rats even have better fur than animals with a normal food intake. It's possible to cut their food too far, which obviously leads to starvation—but get things just right and hungrier rats will significantly outlive members of their species who eat what they like, in significantly better health. These findings show us something remarkable: aging is not some rigid, immutable biological inevitability. A deceptively simple treatment can slow down almost everything about it, all at once, across the animal kingdom.

What seemed for the vast majority of human history to be a fixed fact of nature can in fact be altered by just eating less. What's more, aging seems to be, on some level, a coherent process: these extreme diets don't just prevent a single age-related disease, but all of them at once, at the same time as deferring frailty and death. That means it's not impossible to imagine that we could come up with medicines that could slow down or even reverse "aging" writ large, not just its individual components. Though it wouldn't be so christened for a few more decades, this was the birth of biogerontology: the study of the biology of aging.

With hindsight, the fact that aging is at least somewhat coherent should be obvious. The fact that we start to suffer simultaneously from a diverse family of ailments, each with their own complex underlying causes, should set off scientific alarm bells. The clogged arteries of heart disease, the dying brain cells in dementia and the out-of-control cells of cancer don't seem to have much in common— so why do they all happen at once? It might seem like just a cruel coincidence, if it weren't for our long-lived, hungry rats in which all of these are delayed together: what this suggests is that there is an underlying, ticking clock which, with surprising synchrony, unleashes a phalanx of terrible diseases on our bodies.

The fact that aging is malleable could save and improve billions of lives. The aim of anti-aging medicine is to replicate in people what we've seen at work in so many dietarily restricted species: to keep us fit and disease-free for longer. This is sometimes referred to as increasing "healthspan"—prolonging the period of life lived without disease or disability.

Dietary restriction is just the beginning. After all, when the first results were published in 1935, we didn't know the structure of DNA—in fact, back then we weren't even entirely sure that DNA was the medium of heredity. Nowadays, we can read an organism's entire DNA sequence in a few hours. Our understanding of how life works has expanded exponentially, thanks to an array of biological tools and techniques which would have sounded fantastical a century ago. Our modern understanding of aging biology, like all science, comes from researchers standing on the shoulders of their forebears—and research into aging runs the gamut from ecology to lab biology.

There's inspiration to be drawn from the diversity of life on Earth, involving an ensemble cast of incredible animals which, it turns out, age at remarkably different rates. We've already met negligibly senescent tortoises which have mastered biological immortality. How could they evolve when aging seems universal? Even if we stick to animals more

closely related to us, mammals have lifespans ranging from just a few months for some unfortunate
kinds of rodents to probably centuries in the case of
whales. How has this diversity of lifespans evolved,
and what tricks might these creatures be able to
teach us about how to age well?

Then, there's what we know from the lab. We've
uncovered huge results in tiny nematode worms: a
change in a single gene, indeed a single letter of
DNA, can extend a worm's lifespan tenfold. We've
also had success in animals with physiologies far
closer to our own: we can routinely improve the
aging process in mice via dozens of different treatments. We've discovered drugs which can slow
aging, or turn back the clock entirely, some of which
are already being tried in patients.

This collection of observations and evidence is
tantalizing and foreshadows a future where aging
will be treated. And this future may not be so far
away: in the last decade or two we have finally been
able to say with confidence what aging **is.** And once
you know what something is, you can start to work
in earnest to target it.

We now think that aging isn't a single process,
but a collection of biological changes which make
old organisms different from young ones. These
phenomena impact every part of us—from genes
and molecules to cells and whole systems inside
our body—and go on to cause the aches and pains,
worsening sight, wrinkles and diseases of the elderly.

We are now at a stage where we can draw up a list of these changes and conceive treatments to slow or reverse each of them.

The ideas for treating aging processes aren't pie-in-the-sky theoretical biology—they are being tested in labs and hospitals around the world today. One such phenomenon is the accumulation of aged "senescent" cells in our bodies. Few in number in our youth and accumulating with time, senescent cells are associated with a number of age-related diseases. In 2011, the removal of these cells in mice was shown to defer multiple diseases and extend lifespan. By 2018, drugs which destroy these cells were undergoing clinical trials in people.

The dream of anti-aging medicine is treatments that would identify the root causes of dysfunction as we get older, then slow their progression or reverse them entirely. They would address many conditions at once, and would also be preventative rather than palliative—reducing your odds of getting diseases, and tackling the more everyday symptoms like wrinkles and hair loss at the same time. We wouldn't wait until patients were old and sick to start with treatment, as we do today; instead, we'd give them in advance, and stop people from becoming ill and infirm in the first place.

Treating aging itself rather than individual diseases would be transformative. Much of modern medicine targets symptoms, or at least takes aim at factors several steps removed from the root cause of many illnesses. For example, if someone has high blood pressure (as many people do, especially as they get older), they will often be prescribed medication to lower it. Many common blood pressure drugs work by relaxing the muscles around the arteries, causing them to widen, and allowing blood to flow more freely. This doesn't address the stiffening of the artery walls, or the clogging of their interiors which are actually **causing** the rise in blood pressure. It's not that these treatments are useless—blood pressure is reduced by these pills, and patients live longer as a result—but these drugs and others are work-arounds and, ultimately, they can never be a cure.

Medicine for aging itself could rejuvenate aged vessels and restore blood pressure to safe, youthful levels for the long term—and those same medicines would improve other aspects of our aging physiology, too. The same biological processes which cause our blood vessels to stiffen are behind other problems, from arthritis to wrinkles; fixing root causes helps with many problems at once. Not only that, but truly controlling high blood pressure would go on to reduce the odds of further problems, from kidney disease to dementia, which are caused when blood pressure is high for prolonged periods. The changes which happen in our molecules, cells, organs and bodies as a whole as we age are why we are so susceptible to disability and disease—identify and learn to treat them, and the ill health of later life can be postponed.

Though we've enjoyed significant success by treating diseases separately, this approach can't add much more to life expectancy: even a hypothetical complete success at curing a single disease has a counterintuitively small effect on health. Demographers can use mathematical models to simulate the total eradication of particular diseases and see what happens to life expectancy and disease burden overall. These calculations show that a complete cure for cancer—currently the leading cause of death—would add less than three years to life expectancy. The numbers for runner-up heart disease are similarly slight: two years at best. The reason for this is simple: if cancer or heart disease don't get you, there are plenty of other diseases waiting in the wings to end your life a few months or years later. And curing cancer, heart disease and all the other diseases which are consequences of aging wouldn't address consequences that we don't currently label as diseases—the frailty, the forgetfulness, the loss of independence and so on. Medicine which tackles the underlying causes of aging would lessen both the risk of disease and the other symptoms of growing old.

This will be the greatest revolution in medical care since antibiotics. Penicillin is a single drug, and yet it can be used to treat a wide range of diseases. The same will be true of treatments for aging—but instead of warding off external threats like bacteria, medicine for aging will target our bodies' own internal degeneration with time.

Even if we can't cure aging in time for ourselves, investing in aging research is an investment in future generations. You only have to invent a new drug or treatment once, and then everyone on the planet, and everyone yet to be born, can go on to benefit from it. Cancer, heart disease, stroke, Alzheimer's, infectious diseases, frailty, incontinence and many, many more—progress on any one of which would be a cause for celebration—could be delayed and maybe defeated together. The legacy of our generation could be treatments for aging which will benefit every generation to come. Initiating the scientific and cultural shift necessary to acknowledge that aging as an entity should be treated could well be the single most consequential thing we do.

The consequences will be profound and wide-ranging, for all of us personally, for our friends and families, and for society and humanity as a whole, and the benefits will far outweigh the costs. Many people's initial reaction to the idea of treating aging is cautious, or even hostile: we wonder what the consequences of longer lifespans will be for population growth or the environment; if treatments for aging would primarily benefit the rich and powerful; or whether dictators could live forever, imposing endless totalitarianism. However, almost any objection can be answered by turning the question around and replacing it with a simple hypothetical alternative: if we lived in a society where there was no aging, would you invent aging to solve one of these problems?

Would creating aging and condemning billions to suffering and death be a viable answer to climate change, or global overuse of resources? Surely we'd find other ways to reduce our collective footprint on the planet before resorting to such barbarism. Similarly, invoking aging to limit the reign of even a particularly despotic ruler is a plan which goes far beyond the craziest CIA assassination plots. Looking at things this way around, the answer is clear: aging isn't a morally acceptable solution to any serious problem. This means that the converse is also true: arguing to leave aging intact in an attempt to avoid other problems is misguided when aging itself carries a huge human cost.

If the conclusion that we should try to defeat aging seems strange, I think that's in large part because of our understandable comfort and familiarity with how things are. We literally get a lifetime to come to terms with growing old, and the place we most often imagine longer lifespans is dystopian science fiction. Our sympathy for the status quo blinds us to how powerful the case for curing aging is— as strong as the case against creating it would be if it didn't already exist.

The moral case for treating aging is underlined by a family of conditions known as progerias, from the Greek for "prematurely old." Sufferers experience symptoms of aging at an accelerated pace: patients look old way before their time, with thinning skin and gray hair manifesting as early as childhood in the most severe forms. People born with Hutchinson–Gilford progeria have a life expectancy of 13 years, and usually die of heart disease—a problem which is otherwise unheard of in teenagers. Another related disease, Werner syndrome, sees patients get cataracts and osteoporosis in their twenties and thirties, before dying of a heart attack or cancer at an average age of 54. These conditions are perhaps the best argument for calling aging a disease, and treating it like one—if this collection of problems is labeled a disease when it appears early, how is it different when it happens at what's currently the "normal" time?

I want to convince you that we should unashamedly aim to cure aging. I'm not saying "cure" because I necessarily think it will happen soon, or all at once, but because I want to normalize an idea which can at first sound jarring. The first treatments for aging will slightly increase our healthspans and maybe our lifespans, too, and that's great. But we shouldn't stop there—what we should aim for is negligible senescence: a risk of death, disability, frailty and illness which doesn't depend on how long ago you were born. Our chronological age would no longer be the defining number by which we live our lives—we would, as individuals and as a civilization, be ageless. That is what a real cure for aging would look like, and it's something we could and should aim for as a species.

Curing aging doesn't mean living forever, but it would lead to a substantial decrease in suffering. Lifespans would increase as a side effect, just as they would if we cured cancer, diabetes or HIV, none of which we are ashamed to aim for. If we did **completely** cure aging, all it would mean is a risk of death which stays constant no matter how chronologically old you are, rather like a giant tortoise. You could still die of an infection or in a road accident, meaning that immortality is not on the cards for a while yet (though hopefully longer lives would lead us to be increasingly proactive in reducing deaths from these preventable causes, too). A cure for aging would change what it means to be human—but, at the same time, it's simply a natural extension of the goals of modern medicine.

This is a hugely exciting time to be alive. Removing the aged, senescent cells we just briefly mentioned went from lab experiment to entirely new treatment paradigm in less than a decade. There are many other ideas which slow aging in lab animals that could soon be making the same journey. Most of you reading this book should live long enough to take the first true anti-aging treatment, whatever that turns out to be.

There are literally billions of lives at stake. This is the science that could save them.

PART 1

An Age-Old Problem

The Age of Aging

Cast your mind back 25,000 years. It's late on a warm spring afternoon in what we now call southern France, and you're gathering firewood a short way from your camp. The men are out hunting, carrying spears and seeking out game like deer and bison. You and your fellow nomads look pretty much like modern humans, but life is very different—not least because of the ever-present risk that it will end suddenly.

At 28 years old you're doing quite well for a prehistoric woman. There are risks everywhere. A tiny scratch could get infected and kill you; you could meet a sudden end in an accident or animal attack; or other prehistoric humans, hungry and desperate, could murder you in a fight. Most tragic of all, though, out of the five children you've given birth to, two have died—one shortly after birth from what we'd now understand to be a serious fever, and another aged three who you buried just a month ago. Prehistory is a dangerous place to live, and death seems to strike at random, often without obvious cause. There's no understanding of germs or birth defects—perhaps you blame capricious, vengeful gods or spirits in an attempt to make sense of it all.

It's hard for us to work out exactly how long prehistoric humans lived, not least because prehistory is defined as the period prior to written records. There were no birth certificates or insurance companies compiling detailed mortality tables. However, surveys of bones at a handful of archaeological sites, plus extrapolation from modern hunter–gatherer societies, give us some idea—and it's at once better and worse than you might expect.

First, the bad news: life expectancy was poor, probably somewhere between 30 and 35 years. Statistically speaking, a lot of you reading this would already be dead. However, life expectancy is a number which can obscure as much as it reveals. This is because it's an average and comes with all the attendant statistical pitfalls. The main reason that it was so strikingly low in prehistory was the appalling rate of infant and child mortality. Infections in the first few years of life struck down many, many babies and children. You probably had only a 60 percent chance of making it to 15 years old—barely better than tossing a coin. This huge number of deaths at young ages drags down the average age at death significantly.

However, if your coin came up heads and you made it to your late teens, you could expect to live another 35 or 40 years, taking you comfortably into your fifties. This "remaining life expectancy" is itself an average, so it's quite likely that a few ancient humans did make it into their sixties or seventies—what we, in modern times, would start to call "old age." A headline 35-year life expectancy both masks the terrible toll of childhood deaths and underplays how long the oldest early humans lived. Such is the challenge of summarizing a complex phenomenon like human lifespan with a single number.

This was the story for tens of thousands of years: eye-watering levels of child mortality held overall life expectancy down; most who made it to adulthood lived decent but not exceptionally long lives. For millennia, death was an omnipresent feature of human life, often rapid and without warning. Those who escaped the capricious clutches of infectious disease, injury or bad luck were greeted by an inexplicable state of decline which we'd now recognize as aging: a gradual loss of faculties in a world where fitness, keen senses and mental acuity could be the difference between eating and being eaten.

It can be tempting to think of prehistoric humans as primitive, but their brains were actually very similar to our own; it seems likely that this constant, senseless loss would take its toll. While we can only speculate, there are sites where human or pre-human remains are found together, suggesting some kind of deliberate disposal of the dead. There's an ongoing debate about when exactly funeral rituals arose—many, of course, would leave no trace which could survive the intervening millennia. But, if these sites are what they appear, funeral behavior could date back tens or even hundreds of thousands of years, to a time before our species, **Homo sapiens,** when our hominin ancestors walked, and aged, upon the Earth. As we transition into recorded history, our preoccupation with death is hard to deny: increasingly extravagant structures culminating in the pyramids of ancient Egypt formed the engineering pinnacle of increasingly rich mythologies surrounding life and its end.

Given this, it's perhaps not surprising that some of the earliest philosophers grappled with aging and death. In ancient Greece, Socrates and Epicurus were unworried by dying, believing that it would be like eternal dreamless sleep. Plato was similarly sanguine, but for different reasons: he believed that our immortal souls would go on existing even after our bodies had ceased to. Aristotle was more concerned by death, and arguably the first philosopher to make a serious attempt to explain aging scientifically in 350 BCE. His central thesis was that it was a process by which humans and animals dry out; as you will note from its absence in the rest of this book, sadly his theory hasn't stood the test of time.

Even as schools of philosophy, religions and empires rose and fell, surprisingly little changed about lifespans for thousands of years. A family moving to London in 1800, looking for work as England industrialized, would have a surprisingly similar tale to tell as their nomadic ancestors, statistically at least. The exact causes of death were quite different—fewer hunting accidents, more factory accidents, and a different spectrum of infectious diseases at home in densely populated urban centers rather than small, nomadic groups—but the result was much the same: high birth rates, high death rates. By this time in history we do finally have some actual data to go on—the two countries whose records stretch back the furthest are the UK and Sweden, and both show an overall life expectancy somewhere around forty years in the early nineteenth century.

As the 1800s got into full swing, things finally began to change. Slowly, between 1830 and 1850, graphs of life expectancy start to pull up. If we take the leading country in the world at any given time, which we can regard as the state of the art in population health for any given moment in history, a very striking picture emerges. The maximum global life expectancy has increased by three months every year since 1840, with clockwork regularity. Even better, the trend shows no sign of abating. Predicting the future is always hard, but you could do worse than to extrapolate this trend of nearly two centuries onward into it. That means (if you're middle-aged or younger), for every year you stay alive your expected date of death recedes a few months into the future.* Alternatively, for every day you survive, you gain another six hours—meaning that a good night's sleep isn't really time wasted, as you're getting most of it back thanks to rising life expectancy.

The cumulative effect of this incredible progress is that lives are now twice as long on average as they were in the early 1800s—life expectancy has gone from 40 years back then, to over 80 in the rich world today. It's easy to be glib about this meteoric rise because it's so familiar, so take a moment and transpose those dry numbers onto your own life. At 40, you'd be statistically dead in 1800; now, you've got as much life again still in front of you. A 20-year-old today has better odds of having a living grandmother than a 20-year-old in the 1800s did of having a living **mother.** In a bare couple of centuries—perhaps 0.1 percent of the total timeline of our species—we've already redefined (indeed, doubled) what it means to be human. Families are now multigenerational, we can plan for the long

---

* I'm being deliberately vague here, because life expectancy applies to populations, not individuals: it's not as if some fraction of you is gradually giving up smoking, for example. However, many of the effects are shared and, if you already don't smoke, you're ahead of the whole-population life expectancy anyway, so the rise in population life expectancy still provides a guide.

term under the assumption we'll live to see it, retirement is more than a few years of ill health for a handful that make it that far. For the first time in human history, the majority of babies born today will have the chance to grow old.

The straightness of the line showing life expectancy increasing is almost suspicious, because those improvements are underpinned by a tangle of cultural shifts, public health measures and scientific and medical breakthroughs happening more or less at random. And yet, every year: three more months. Successive phases of this revolution have been driven by very different phenomena. It started with the taming of humanity's greatest ancient foe: infectious disease.

Pandemics are a humbling reminder of the power of nature compared to our own. The coronavirus crisis has laid bare what many of us had largely forgotten: the terrible toll infectious disease can take without treatments or vaccines. Nonetheless, your risk of death from COVID-19 is still substantially less than the risk from infections in the past—taken over the whole of human history, bacteria, viruses and other microorganisms have probably struck down more of us than anything else. Even in the worst case, the consequences of coronavirus are unlikely to exceed those of the 1918 flu pandemic: during that outbreak, 50 to 100 million died over a couple of years at the hands of flu viruses—up to 5 percent of the global population at the time—dwarfing the 20 million killed in the preceding four years of mechanized annihilation in the First World War. Humanity would do well to remember that our real enemy is not one another.

However, through the 1800s, unhygienic towns and cities were refitted, open sewers replaced, public health initiatives began to take root and infectious disease began to decrease. Science and medicine entered the fray, first with vaccines, and then germ theory, demonstrating that it was tiny, invisible organisms—not bad air or bad luck—that drove infection. Vaccination has since wiped smallpox from the face of the planet (albeit shockingly recently, in 1977), is well on the way to doing the same for polio, and has made former childhood specters like diphtheria and whooping cough so infrequent as to sound antique. Improved fertilizers and mechanization of agriculture ushered in better nutrition population-wide, leading to healthier children and adults better able to fend off many causes of death, infections included. At the same time, the twin engines of education and economic growth were lifting millions out of poverty, compounding the improvements in food and cleanliness. Better health and increasing longevity also bolstered the economy, in a virtuous circle of burgeoning health.

In 1850, life expectancy at birth was around 45 years in chart-topping Norway. By 1950, Norwegians (who reclaimed the crown after nearly a century of dominance by New Zealand) could expect to live beyond 70. Progress was largely driven by improvements in early and mid-life. Infectious diseases disproportionately struck children, but were also prevalent in adulthood—and their reduction dramatically increased life expectancy overall.

It's in the last seventy years or so that improved life expectancy at older ages has finally started to move the needle on life expectancy overall, mainly due to huge strides in medical science and healthcare provision, and healthier lifestyles. Taking a survey of other essentials of modern medicine—automatic defibrillators, stents, dedicated coronary care units in hospitals, heart bypass surgery—you'll find that **none of these** was available in 1950. Even cardio-pulmonary resuscitation, or CPR, which uses chest compressions to restart a stopped heart and is a staple of TV drama to the point of cliché, hadn't been invented yet. Also missing were preventative drugs like statins which lower cholesterol and make heart disease less likely to happen in the first place. And all of this is just in the world of cardiology. Drugs, devices and surgical techniques have improved outcomes for people with many different diseases at all ages, but their effect has been particularly important for the survival chances of older people. This is because, since we have massively reduced infectious disease, the deadliest health problems today are things like heart disease and cancer, which primarily hit later in life.

Of the improvements in lifestyle, the single

biggest has been the decline in smoking. It's shocking, but the shadow of a single industry—indeed, by and large a single product, the cigarette—is cast over life expectancy statistics across half a century. In the U.S. in the 1960s, cigarette consumption was high enough for every adult to smoke more than half a pack a day. This generation of long-term smokers created a population-wide crescendo of smoking-related diseases which, because it takes time for smoking to cause disease and death, peaked a few decades later in the 1980s and 1990s—at which time around a sixth of all deaths (and a staggering 25 percent of male deaths) in the developed world were attributable to tobacco. In total, there are estimated to have been 100 million deaths from smoking in the twentieth century. Smoking rates have more than halved since their peak, and are still dropping—and that drop is now showing up in life expectancy statistics.

The cumulative result of all of this can be seen in global tables of life expectancy: the country with the highest life expectancy in 2019 was Japan, whose citizens live to 84.5 years old on average. And there are plenty of others snapping at its heels—the top 30 countries in the global rankings all have life expectancies above 80 years.

As well as extending lifespan, we've also been extending healthspan. A study looking at changes in the UK between 1991 and 2011 found that life expectancy at age 65 had risen by about four years, and so had the number of years spent without cognitive impairment, and, if you ask people to rate their own health in a survey, the number of years spent healthy went up by a similar amount again. Improvements in health are most pronounced in the very old: the fraction of over-85s in the U.S. classified as disabled dropped by a third between 1982 and 2005, while the number who are institutionalized almost halved over the same period, from 27 to 16 percent. Depending on how you measure health or disability, the fraction of our lives spent in ill health is either shrinking or roughly constant, both of which are good news.

The only wrinkle in the data is that, while severe disability is dropping, minor disability—conditions like arthritis which are painful and inconvenient but, except in very advanced cases, still allow sufferers to go about their daily lives without help—seems to be on the rise. One issue might be improvements in diagnosis and recording of diseases and disabilities rather than a true increase in their prevalence. Early detection of diseases can have complicated effects: on the one hand, people seem to get more diseases at younger ages when we look at statistics; on the other, medical or social care can often intervene earlier and improve and extend lives. There is also substantial variation in healthspan between countries, but since it's much harder to pin down than lifespan there's room to debate what exactly underlies these differences.

This isn't a picture totally free from qualification

and nuance, but it's substantially more positive than stereotypes about medical care merely prolonging our years in frailty would have you believe. And, purely theoretically, this is what we'd expect: to die, you have to die **of** something, which implies you'll be sick; and the reverse is also true, with diseases which cause significant disability, like heart disease and dementia, also being deadly. It would be strange indeed if we had substantially extended life without postponing disability, so it makes sense that, broadly speaking, that isn't what's happening.

Our story so far has concentrated on the rich world. What about in less wealthy countries? The answer, since 1950 at least, is good news. Low- and middle-income countries are very rapidly catching up with those whose historical good fortune allowed them to lead the way. The prevailing tale in developing countries has been one of meteoric rises since 1950: life expectancy in India has almost doubled, from 36 in 1950 to 69 today. The result has been a dramatic reduction in health inequality in the last century. Even in 1950, there was a stark separation between rich and poor: while life expectancy in India was 36, Norwegians could expect to live to 72. Today, Indians are only 10 or 15 years behind the chart-topping countries. Overall, 90 percent of the world's population now lives in countries where life expectancy is over 65 years, and 99 percent in countries where it's over 60. Though we, of course, have a moral duty to help those in countries stuck with low life expectancies, thankfully, and in a difference from just 50 years ago, they are the exception rather than half of the world. The end result of progress over the last two centuries is that most of the world's population is most of the way through the story just told.

This means that now, uniquely in human history, we have become victims of our own success—the rout of malevolent microbes, the rise of public health, healthier lifestyles, modern medicine and increasing education and wealth have conspired to bring us face-to-face with a new scourge: aging. No matter where you live in the world, you're very likely to live long enough to experience the frailty, loss of independence and diseases associated with getting old. This is the age of aging.

The age of aging is a strange time to be alive, but it's hard for us to appreciate because we all live in it. Most lives have a fairly similar, well-defined structure, and this universality conceals how dramatically different this is from life even a century ago. Though some people's years are cut short by tragic accidents or disease, they are the exception: most will get to enjoy the classic three-stage life which we've become used to—education, then work, then retirement.

That structure is geared to the length and shape of a human life—just not necessarily the lives we live today, or will in the near future. We spend our first couple of decades in education, not thanks to some dispassionate analysis of the optimal duration of learning and development, but because we need to hurry on to the next stage and start work. Then we try to earn money for 40 or 50 years, partly to provide for ourselves, partly to pay taxes and help the next generation through their early years and assist those already older than us, and partly to save for our own old age. Careers reflect this, with a steady rise through the ranks until we are in our forties or fifties, and a winding down thereafter. The duration and nature of this period isn't optimized either, but an accident of history, with "retirement age" pegged to the age of onset of serious ill health during the first half of the twentieth century.

It's tempting for those alive today to assume that the three-stage life, divided approximately as it is now, has been the norm for far longer than is actually the case. In fact, even 50 years ago, far fewer people were alive and in good enough health to enjoy retirement at all. Thanks to rising life expectancies and falling birth rates worldwide, between 1960 and 2020, the global population aged over 65 grew dramatically faster than the population overall: it almost quintupled, from 150 million to 700 million people; by 2050, it's predicted to more than double again, to 1.5 billion—meaning that over-65s will make up one in six of the global population. The older the slice of the population you look at, the more rapidly it's growing: the number of people aged 100 or over (centenarians) has gone from 20,000 in 1960, to half a million today, to a projected three million in 2050—a change of more than a hundredfold in less than a century. And, like life expectancy, population aging is also moving faster in the developing world than it did in the rich countries: France, the U.S. and the UK took 115 years, 69 years and 45 years respectively for the fraction of the population aged over 60 to double from 7 to 14 percent; projections for Brazil suggest it will undergo the same transition in just 25 years. This means that poorer countries will have even less time to adapt to the coming silver tsunami.

The social and economic impact of this age of aging will be dramatic if we don't act quickly. Pensions provide a salutary and simple example. The first state pension in the UK was paid in 1909 to people aged over 70, and the scheme was updated in 1925, reducing the pensionable age to 65. In 1948, the state pension was made universal, alongside reducing the pension age for women to 60. The pensionable age for women wasn't changed until 2010, at which point a sliding increase was implemented to make the starting ages for men and women equal, in line with equality legislation. Men's pensionable age finally rose in December 2018, meaning that the age at which men received a state pension in the UK remained constant for almost a century—during which time, life expectancy in the UK increased by 23 years. That successive governments sat by as life expectancy climbed, leaving what is now quite predictably one of the biggest items of government spending largely untouched, is mind-blowing. The simple fact is, we will need to work longer to pay for a larger fraction of our life spent retired.

We've been fortunate that decades of economic and population growth mean that pensions haven't become a crisis before now, but a crisis is approaching if we do nothing. This news also has a rather positive spin which is rarely highlighted—because we are living in good health for far longer, many current 65-year-olds are more capable of working than our forebears were at the same age. This gives us more time to contribute to the economy and fund our retirement—which could still be longer, healthier and wealthier than in the past. A 65-year-old was old in the 1920s—only just over half of people made it to that age, making 65 back then roughly equivalent to being in your early eighties today. While increasing the retirement age to 80 may raise some eyebrows, there's clearly some headroom above 65 where a compromise could be reached.

More broadly, the age of aging highlights the need to reinvent the three-stage life as lives extend. Lifelong education and training will become increasingly important. A life which starts with 20 years of training and ends with 20 years of retirement means a 40-year career when life expectancy is 80; if you live to 100 with the same template, your career is half as long again. Six decades is a long time in the same job—long enough for the job to cease to exist, or for you to get pretty bored of it. No longer will a 50-something be entering the later stages of their career, passing the years to retirement; they could instead take a few years off, retrain and start an entirely new career with the productive decades ahead. With both careers and retirements extending, perhaps we won't want to work for decades and then retire for decades more, but instead take periodic sabbaticals to return to education, travel or take up new hobbies at different times in our lives. It seems unlikely that three stages is the most efficient life structure even now—let alone as we continue to live longer.

Another feature of the age of aging is the significant fraction of our resources we devote to taking care of the elderly. Thanks to their multiple diagnoses and medications, the average 80-year-old costs the healthcare systems in the U.S. and the UK about five times as much as the average 30-year-old. This is another way in which aging has been internalized, even industrialized, in our societies. Hospitals, care homes, nurses, doctors, administrators, pharmaceutical companies, medical device manufacturers and more besides make up a system that swallows up a substantial fraction of our economies—typical rich countries like the UK and Germany spend roughly 10 percent of GDP on healthcare, while the U.S. dedicates an even more substantial 17 percent—thanks in significant part to the chronic diseases of aging. The growing need for long-term medication and care for the elderly means that these fractions are forecast to increase.

On top of these "direct costs" of treating the diseases of aging, there are also "indirect costs," like people giving up work due to chronic illness, or cutting back their hours to take care of a friend or relative with one. These are often hidden from view and politically neglected, but the indirect costs of diseases like cancer and dementia frequently exceed the direct ones. The total cost is enormous: unpaid care alone in the UK is estimated to be worth about the same as the entire healthcare budget. Once again, this was never planned—but love and a sense of responsibility make up where official provision falls short. We tacitly accept the enormous burden on spouses, children and neighbors and, as gradually more of us get old and unwell enough to need support, this unofficial system will be strained even more than it already is.

As the age of aging comes into its prime, these costs are going to become increasingly unsustainable. Alongside frank discussions with voters about pension provision, health and social care, our long-term strategy needs to include research into medical treatments for the aging process itself.

What's astounding is that the doubling of human life expectancy since the start of the 1800s has been achieved without any treatments for aging. We've scored some indirect hits—improved diets, exercise, cutting out smoking, and preventative medicines to reduce cholesterol or blood pressure all arguably slow parts of the aging process to some extent—but there's not a single drug or treatment available in your local pharmacy or hospital expressly designed to slow or reverse aging.

In fact, regulators around the world—like the Food and Drug Administration in the U.S., or the European Medicines Agency—wouldn't grant a license to sell an anti-aging treatment even if it did exist. Drugs will only be approved if they treat a specific disease, and aging is not recognized as a disease, but dismissed as a natural process. This might sound like an insurmountable obstacle to treatments for aging, but scientists are in the process of overturning these rules—we'll talk about how in Chapter 11. And there are some hints that aging is starting to be recognized: in 2018, the World Health Organization added a new code to the International Classification of Diseases: XT9T, for conditions that are "aging-related." The scientists who proposed its inclusion hope it will smooth the path to developing therapies.

Even if we continue to consider aging an immutable fact of life, it seems likely that life expectancies around the world will continue to rise. There are still improvements to be eked out: earlier detection and better treatments for cancer and heart disease can deliver us a few years, even if they stop short of a complete cure; continued improvements in lifestyle, along with better and more universal healthcare provision, can surely add at least a few years too. Given that sublime complexity has led to surprising simplicity thus far, you could do worse than simply extrapolating from the existing trend of three months per year, every year. Running with this assumption makes projections which seem dramatic to contemporary eyes: life expectancies would leap by another 25 years over the course of a century, meaning that we'd expect most babies born globally since the year 2000 to celebrate their 80th birthdays—and most babies lucky enough to be born this millennium in the rich world to celebrate their 100th.

Official projections and many demographers often hold that some intrinsic limit to human life expectancy will eventually stop this increase in its tracks—but no particular reason is advanced for this, and pessimists have been repeatedly proven wrong in the past. One study examined 14 predictions of a limit on human life expectancy, and wryly observed that the average time between a limit being proposed and being broken was just five years.

There are some headwinds to rising life expectancy which could slow progress. One example is the increasing prevalence of obesity. Our growing waistlines are already having a negative impact on life expectancies around the world, but larger positive changes have thankfully been able to outweigh this effect so far. Nonetheless, improving diets and making exercise easier to integrate into daily lives both need to be priorities if we want to make sure lifespans continue to increase. Other factors, from air pollution (whose risks are only beginning to be understood, but seem to affect aging to some extent—not just in the respiratory system, but by promoting cardiovascular diseases and perhaps even dementia), to antibiotic resistance and emerging diseases like coronavirus (which could see a partial return to the bad old days of deaths from infection) are also worth trying to get ahead of. There are also inequalities which mean, while headline life expectancies increase or at worst stay constant in whole countries, some socioeconomic groups or regions have experienced **declines** in lifespan in the last decade or so. But nonetheless, if we keep looming threats at bay, carry on capitalizing on marginal gains and work to ensure that this success is shared, most people in the world having a solid chance of living to 100 by 2100 doesn't sound implausible.

The recent history of life expectancy is perhaps the crowning achievement of humanity. No other scientific or technological advance can claim to have bettered the lives of billions in such a fundamental way.

Living at a time when one singular cause, aging, is responsible for so much—everything from the shape of our life courses, to economies and many of our institutions, and the majority of human suffering and death—is humbling, but also genuinely exciting: by tackling this root cause, science allows us the possibility of doing something about all of this at once.

In order to end the age of aging, we need to

understand what the aging process **is.** Then we can start to consider the treatments which could address it. Therefore, the next few chapters will be spent exploring aging and demystifying this process. Science is finally beginning to understand its components, and identify the surprisingly small number of processes which cause us all to get older. We'll see how breakthroughs in aging biology took the field from a weird, fringe pursuit of theorists, pioneers and cranks to a legitimate, mainstream area of biology.

The best place to start is to look at the near-universal phenomenon of aging through perhaps the one truly universal principle in biology: evolution.

2

On the Origin of Aging

When Darwin visited the Galápagos in 1835, his remit was far broader than tortoise collection. During his stay on the islands, he made meticulous notes on their flora and fauna, as he did at the many other stops during the nearly five years he spent voyaging on HMS **Beagle.** These observations formed part of the huge body of work which underpinned one of the greatest discoveries in the history of scientific thought: the theory of evolution by natural selection.

Darwin published this groundbreaking idea in **On the Origin of Species,** two decades after his stay in the Galápagos. His great insight (which was independently conceived by his contemporary Alfred Russel Wallace) was that animals, plants and all life forms are optimized for their environment by "descent with modification." Baby animals will differ, quite at random, from their parents; the majority of those differences will be negative or neutral, but those handful with advantageous differences will have greater success surviving, reproducing and thus passing on those qualities to their (more numerous) offspring; those offspring will themselves differ in small, random ways, some slightly better and some slightly worse, and so on. Gradually, over successive generations, the best-adapted will succeed over their peers—the so-called "survival of the fittest."

A famous illustration of this is "Darwin's finches," a collection of finch species found on the Galápagos Islands which display remarkable diversity in the shape of their beaks. Darwin observed that, in spite of their diversity, they all bear "the unmistakable stamp of the American continent"—the closest large land mass to the islands. Species which live in different places nonetheless sharing features suggests that they may have descended from a common ancestor, but evolved novel adaptations for their new environment. Thorough work on the finches a century after Darwin's visit finally pieced together the reason for their extreme differentiation: food. Each island offered the birds subtly different food sources—a large beak might give its owner the strength to crunch on seeds, while a pointed one permits catching insects hiding between leaves. Starting from a common ancestor with a single size of beak but some variation between individuals, finches with slightly bigger or smaller beaks could better exploit the local cuisine and pass on their genes. Over successive generations, finches with a beak closer to the optimal shape and size to eat whatever food was available on their island would prosper, ultimately leading to the incredible variety we see today.

Over a century after Darwin's science-shattering tome, evolutionary biologist Theodosius Dobzhansky published an essay entitled "Nothing in Biology Makes Sense Except in the Light of Evolution." This title pithily encapsulates the universality of Darwin's theory. If a scientist somewhere uncovers some fact about biology but it doesn't fit with evolution, they're going to have to rethink it. The alternative would be to rework the whole of modern scientific thought to sideline the most fundamental law of biology. There are so many lines of evidence, theoretical and practical, and so much of modern biology which **does** make sense in light of evolution, that it would take some truly extraordinary evidence to overturn it.

As we've seen, aging is a phenomenon which has stalked humans since the dawn of our species. We also see aging in our pets: dogs with arthritis, no longer so excited to chase a stick; half-deaf cats with eyes clouded by cataracts. Rather more quickly than us, our companion animals too succumb to aging. So do farm animals and, as the study of different creatures, plants and eventually microscopic organisms began to explore all the kingdoms of life, we've found that aging is (almost) everywhere. From mammals like us, to insects, plants and even single-celled organisms like yeast, aging seems to be a near-universal process of degeneration. And this comes as no great surprise—outside of biology, machines wear out and break with time, buildings crumble and fall. Why should living things be any different?

The question is, how can we square aging with evolution? If evolution is about the survival of the fittest, what exactly is fitness-optimizing about a process of progressive degeneration? The other big question is why we see such diversity when it comes to aging. The shortest-lived adult insect is a type of mayfly whose females emerge, mate, lay eggs and die in less than five minutes; the longest-lived vertebrate (animals with a backbone, like us) is the Greenland shark, with the oldest known female estimated to be 400 years old. Why does a mouse live for months, a chimpanzee for decades and some whales for hundreds of years? If aging is a process of wearing out, why do animals do so over such vastly different timescales?

The evolution of aging sounds like a paradox. Thankfully, though, we can make sense of aging in light of and not in spite of evolution. Understanding this is not just an exercise in evolutionary theory (though that aspect is conceptually fascinating), or reconciling two huge and apparently contra-dictory laws of biology (though that is obviously important)—it gives us insight into what aging is, what it isn't, and how therefore we might go about treating it.

We need to start by defining what we mean by aging. We'll start not with a biological definition of aging, but with a statistical one: aging is an increasing risk of death over time. An animal, plant or other life form whose risk of death increases as it gets older can be said to age; a creature whose risk of death remains constant, like the Galápagos tortoise, doesn't. We've already seen that humans' risk of death doubles every eight years—this defines our rate of aging from a statistical perspective. We can use this definition to make sense of aging on an evolutionary level, and everything from wrinkles to risk of heart disease will follow from that.

Starting at the most basic level, people sometimes invoke not biology, but physics to explain aging. "It's just the second law of thermodynamics," goes the argument, "which says that entropy tends to increase"—in other words, things become more disordered, and fall apart with time. All good things must come to a messy, high-entropy end, whether they're steam engines, universes, or animals. This argument is flawed because it omits a crucial phrase: the second law only applies **in a closed system.** If you're isolated from your environment, there's nothing you can do except put off the inevitable descent into decay—but, if you're not isolated, you can import energy from your surroundings and use that energy to power a spring clean. This might sound esoteric, but it's actually quite simple—because animals can get energy by eating, and plants can turn sunlight into food, they are then free to use this energy for all manner of biological and biochemical processes which either recycle, remove or replace critical components which are deteriorating. Life is therefore under no thermodynamic obligation to age.

Unfettered by oversimplified thermodynamics, animals have evolved incredible capacities for self-repair—some, like salamanders, can lose a limb and just regrow it. It's quite the party trick—but there are equally impressive (if less visually striking) phenomena buzzing away on a microscopic scale inside every living thing, including you, all the time. As cells, cellular components or the molecules they are made of are damaged or fall apart, our bodies clear up the detritus and make new, pristine replacements. Myriad molecular machines are constantly maintaining complex structures, ridding our cells of rubbish and generally preserving our integrity. In humans, these processes continue for decades without breaking down. Supplied with energy, there's no reason in principle why these processes should lose their efficacy over time. Why doesn't evolution keep on ramping up the effectiveness of self-repair until it becomes flawless indefinitely?

It was probably Alfred Russel Wallace who came up with the first evolutionary theory of aging. He suggested, in notes written sometime between 1865 and 1870, that older animals "as consumers of nourishment . . . are an injury to their successors": in an environment where food is limited, too many old animals hanging around and consuming resources would make it harder for their descendants to survive. "Natural selection," concluded Wallace, "therefore weeds them out." Animals with a biological expiry date were fitter because they gave their children the space to thrive and have offspring of their own. Independently, a biologist called August Weismann came up with basically the same theory, proposing that lifespan was limited by "the needs of the species."

This theory—and any other which invokes the good of the species over that of the individual—has a fatal flaw. It's an argument based on what we now call "group selection," where an animal acts in the best interests of a group—usually its whole species—rather than following its own selfish motivations. This is problematic because group selection necessitates an uneasy truce. As long as every animal shares a predilection for growing old for the good of the species, everyone wins, but as soon as one is born with genes for slightly longer life, the delicate balance is destroyed. The "selfish" animal would outcompete the altruists: while they all die, freeing up resources for others, it would consume those resources, allowing it to live a little longer—perhaps just long enough to have an additional descendant before dying itself. This extra descendant makes the gene for longer life a little more common in the population and, eventually, animals with this selfish gene for longevity would come to dominate. Repeat this down the generations with ever-more selfish variants, living longer and outperforming one another to an ever-greater extent as time progresses, and aging is no longer an evolutionary advantage—indeed, it's actively selected against, even if longer lives of individual animals are detrimental to the population as a whole.

Group selection has largely fallen out of favor in modern evolutionary biology because this scenario recurs regardless of the trait you choose. Selfish genes will (almost) always create selfish creatures who take advantage of their genetically altruistic peers, eventually coming to dominate.

Instead, rather than invoking some noble utilitarian calculus for the good of the species, we now think that aging evolved thanks not to evolution's intention, but to natural selection's neglect. This evolutionary oversight is an inevitable result of the risk of death at the hands of things like infectious diseases, predators, or just falling off a cliff, all of which are external to the animal itself. Together, these are known as "extrinsic" mortality, as opposed to "intrinsic" mortality, driven by something like cancer which is the result of the animal's own body going wrong. Evolutionary biologists in the middle of the twentieth century realized the significance of extrinsic mortality and laid the foundations of our modern understanding of how aging evolved.

Let's imagine some animals living on an island. Life on the island is perilous—there's a 10 percent rate of extrinsic mortality every year, thanks to predators and endemic disease. That means that, every year, 10 percent of the animals die—they have a 90 percent chance of making it to their first birthday, an 81 percent chance of making it to their second . . . but only 35 percent of them will see ten, and fewer than 1 percent will reach fifty years old. Even though you are less likely to find older animals, there's still no actual aging in this scenario: remember, our definition of aging is a risk of death which increases with time, and the risk of death here is a constant 10 percent. The intrinsic mortality of our animals is zero, however long ago they were born.

Though evolution is often called "survival of the fittest," there is something it cares about far more than survival: reproduction. Evolution's bucket list for an organism contains just one item—have babies. Mutations which make having children more likely mean that a given animal will, on average, have more of them, and the offspring will carry the mutation which helps with reproduction too; over repeated generations, they will reproduce more than animals without the mutation, and gradually come to dominate in the population.

So, back to the risky island: let's consider reproduction there. Even if the animals can reproduce throughout their lives, by far the majority of reproduction is going to occur at younger ages simply because most animals will die before they reach a grand old one. Because most reproduction happens in youth, changes that affect an animal's reproductive chances at older ages won't make much difference. An animal with some modification that doubled its baby-making capacity at 50 years old wouldn't have an evolutionary edge because it likely wouldn't survive long enough to put its baby-doubling powers to use. By contrast, an animal which got a baby bonus aged three would probably still be alive after three years, raring to reproduce, and this trait would mean far more offspring— conferring a significant evolutionary advantage.

The increased ability to reproduce could manifest in lots of different ways, from a literal ability to give birth to larger or more frequent litters, to a longer beak to gather more food and support more children, or just an improved ability to survive long enough to have more children. However it chooses to do it, evolution's power to tweak and optimize in young animals is significant, because they are likely to be alive and able to pass their genes on to the next generation. By contrast, evolution has trouble improving the lot of older animals because they are very unlikely to pass on their genes simply because they won't make it that far. This is the fundamental reason for aging—evolution's inability to keep old animals fit because they are less likely to have children. All of this, remember, is still possible without invoking aging itself—there are fewer older animals purely because of **extrinsic** mortality. Thus, somewhat counterintuitively, the key driver of the evolution of aging is the risk that an animal might die of something other than aging.

The next question is how in practical terms this evolutionary neglect manifests. The first mechanism is "mutation accumulation theory." Mutations are changes to the genetic code, alterations to the DNA which provides the instruction manual for building and maintaining an animal. We are all mutants: though your DNA is a 50:50 mix of the DNA from your mother and father, we each carry 50 to 100 variations which are present in neither Mom's nor Dad's DNA. Most of these variations don't have any effect—they fall into parts of our DNA which don't really make a difference to our survival chances. A handful will either be positive or negative: those which are positive will improve chances of survival or reproduction, and stand a chance of being passed on with greater frequency in the next generation; those which are negative do the opposite, and should be weeded out by evolution over time.

Back to aging, and mutation accumulation theory. Imagine some mutation crops up which makes animals spontaneously die at the age of 50. This is unambiguously disadvantageous—but only very slightly so. More than 99 percent of animals carrying this mutation will never experience its ill effects because they will die before it has a chance to act. This means that it's pretty likely to remain in the population—not because it's good, but because the "force of natural selection" at such advanced ages is not strong enough to get rid of it. Conversely, if a mutation killed the animals at two years, striking them down when many could reasonably expect to still be alive and producing children, evolution would get rid of it very promptly: animals with the mutation would soon be outcompeted by those fortunate enough not to have it, because the force of natural selection is powerful in the years up to and including reproductive age.

Thus, problematic mutations can accumulate, just so long as they only affect animals after they're old enough to have reproduced. Aging under this theory is not something which makes animals fitter—it's just that evolution hasn't got the power to do anything about it. The textbook example of this, which was actually the inspiration which nudged mathematical biology polymath J. B. S. Haldane to come up with the idea of a "force" of natural selection which declines with age in the first place, is Huntington's disease.

Huntington's is a brain disease which is caused by a mistake in a single gene, typically giving rise to symptoms sometime between the ages of 30 and 50, and is usually deadly around 15 to 20 years after diagnosis. As we've seen, human life expectancy in prehistory was somewhere around 30 or 35 years, so, from an evolutionary perspective at least, coming down with Huntington's at age 40 and dying of it at 55 doesn't much matter. A "wild" human was already likely to have had several children, and their remaining reproductive lifespan was short. Even in modern times, it's quite possible for a Huntington's sufferer to have children before succumbing to the disease. Thus, in spite of its deadliness, Huntington's persists, albeit rarely, in the human population.

Huntington's is a clear example of an accidentally accumulated mutation, where a single gene causes something unambiguously and severely bad at post-reproductive ages. But, while deadly single-gene conditions provide clear examples, the bigger deal when it comes to normal aging is the cumulative effect of many different genes, working alone or in combination, conspiring to slightly erode our chances as we enter our post-reproductive years. Deadly mutations float around in the gene pool, to evolution's studied indifference, as long as they kill us slowly enough to allow us to reproduce first. Taken together, these imperfect genes ignored by evolution are behind some of the processes which cause us to age.

Aging isn't purely accidental, though. On top of being indifferent about your post-reproductive well-being, evolution will happily do something even crueller: it will trade your future health for increased reproduction. Evolution will trade literally anything for increased reproductive success.* It will happily trade running speed, height, fur color, or anything else for more offspring per lifetime. If being faster, slower, taller, shorter, darker, lighter, longer- or shorter-lived will improve reproductive success overall, evolution will take it.

So how does evolution make this deal with death and bring on animals' decline in exchange for reproductive success? The answer is that genes often have multiple personalities. Modern genetics tells us that genes don't exist in splendid isolation, coding for a single characteristic. They have multiple functions at different times and in different parts of the body, and interact with one another in complex networks. You should also raise an eyebrow if you ever hear someone speak of a "gene for" a complex characteristic. Even traits as simple as eye color are under the control of many different genes, and those genes also have multiple functions, playing a role in hair and skin coloration, and they may well moonlight in other processes in ways we are yet to uncover. The

---

* This is actually a circular statement. We see the survivors of millions of generations of living things, and the reason that we see the organisms we do is that they were more reproductively successful than those who didn't make it. It's often easy to talk of it as though evolution itself is an entity, "wanting" this, "making trade-offs" for that, and so on (and I will continue to do so unapologetically throughout this book!). But really, "evolution" is just a word for the passive, tautologous process of things with high reproductive fitness being good at reproducing.

multifunctionality of single genes is known in biology as "pleiotropy."

Thus, the second idea in the evolution of aging is
"antagonistic pleiotropy"—the idea of genes which
have multiple effects allowing them to conspire to
aid reproduction in early life, but go on to cause
problems as the animal gets older. Imagine a mutation in our island-dwelling animals which increases
the risk of an animal dying over the age of 30, but
which allows it to reach reproductive maturity a
year sooner than otherwise. Carriers of this mutation will rapidly expand in numbers compared to
those without—the disadvantage to the few percent
of animals left alive post-30 is dwarfed by the enormous reproductive advantage which will accrue to
young animals who now have an extra year to reproduce at a time when most are still alive.

So not only can mutations whose negative effects
occur late in life accumulate by accident as in mutation accumulation theory but, if such mutations
have a positive effect on reproduction overall, they
will be actively selected **for.** How many years in your
eighties would you give up for better physiology as
a young adult? Evolution can answer that question,
ignoring any poetic musings on the folly of youth
and wisdom of age, and instead optimizing across
the generations to maximize reproductive success.

The behavior of these antagonistically pleiotropic genes is a little abstract, though. Why would
reaching reproductive maturity more quickly result

in an earlier demise? Making things slightly more concrete introduces our third and final evolutionary theory of aging, known as "disposable soma theory." It arises from a principle which will serve you well whatever trait you're trying to explain the evolution of—in nature, as in everyday life, there is rarely such a thing as a free lunch. Recall how we demolished the thermodynamic argument against aging: animals and plants can acquire energy from their environment, and use that energy for repair and maintenance. Physics says we don't have to grow old, just as long as we're willing to spend some of our energy, hard-earned by long hours hunting and gathering food, warding off the ravages of time and entropy.

Immortality, in biology as in mythology, always comes at a price. In biology, it's not an ironic forfeit to amuse the gods, but the necessity to maintain your body for an indefinite period of time. Maintenance costs energy—energy which could be used growing muscles to outrun predators, developing an immune system to fend off disease, or becoming reproductively mature more quickly and producing offspring before something kills you.

Disposable soma theory takes this notion of allocating limited energy between different tasks and applies it to reproduction and aging. The soma is biologists' word for the cells of the body, as opposed to reproductive cells like eggs and sperm. While it's perhaps a little depressing to look at yourself this way, as far as evolution is concerned you're just a vessel for either sperm or babies. It's been the mantra of this chapter so far: evolutionary success and reproductive success are synonymous. Your children are important, but your body, or soma, is expendable. That means that looking after those re-productive cells is of paramount importance, and all creatures are going to expend energy keeping them in top condition. However, it's less clear how much energy to spend on maintenance of somatic cells. As with the preceding theories, all evolution really cares about is that you're capable of sticking around long enough to pass on your genes.

So, as a creature with a limited amount of energy to spend, would evolution rather you spent it on maintaining a pristine physique into your dotage, or gearing up rapidly for reproduction? Evolution will do the sums depending on the level of extrinsic mortality. If it's reasonably high, evolution is often going to favor the latter, making sure you've had children who'll outlive you and leaving your disposable body to break down with age (if you even survive long enough for that to happen). So, one way that antagonistic pleiotropy could act is via mutations which imbue a lackadaisical approach to somatic maintenance which allows you to grow more quickly as a youth, but which will come back to haunt you when your imperfect body, thrown together in a hurry, reaches old age.

The best way to see these theories in action is to

look at the incredibly varied lifespan and reproductive strategies of different animals. Given the intimate relationship between the evolution of aging and extrinsic mortality, we might expect animals who live in more hazardous environments to reproduce rapidly, and age more quickly once they have. We can see how this holds up by considering two extremes of mammal longevity: mice and whales.

Mice live in a very high-risk environment and need to expend a lot of energy on two things: evading the sharp eyes and claws of cats, and producing numerous descendants quickly before they get ill or get eaten. This means they don't have a lot of energy left over for optimally maintaining their somatic cells. This tallies with what we see: mice have litters of six to eight pups, and can breed once a month; and they typically live less than two years in the wild. In the benign environment of the lab, they can survive three or four years before succumbing to old age—substantially longer than in nature, but still 20 or 25 times less long than humans.

If instead you're a whale, monarch of the oceans, and subject to few natural threats, you can afford to chill out, mature more sedately and have children at a leisurely pace. This pushes back the date at which it's evolutionarily acceptable to die at the hands of accumulated mutations or genes which were helpful in youth, and makes it far more biologically worthwhile to put plenty of energy into somatic maintenance. Accordingly, whales are some of the longest-lived mammals: the record holder is the bowhead whale, with one male found in the wild estimated to be 211 years old.* Bowheads don't become reproductively mature until they're in their twenties, and typically have one baby at a time, every four or five years.

Aging a whale is tough, and the 211-year record was calculated by chemical analysis of the lenses in the creature's eye—but a remarkable tale of a whale that got away gives us direct proof of bowheads' exceptional longevity. In 2007, Inuit whalers (one of the few groups permitted to continue whaling on a subsistence basis) caught a bowhead whale which had a harpoon lodged in its bones. The weapon was identified as a type of "bomb lance"—a horrific device designed to explode a couple of seconds after spearing its mark—patented in 1879. Unless the harpoon was an antique at the time of use, this puts the whale's age at significantly over a century—and to that we need to add on the fact that the whale was big enough not only to be worth hunting, but to shrug off the assault. We might even be underestimating the true maximum lifespan of bowhead whales. For starters, we've not actually checked the age of all that many of them, so there could be much older ones swimming around which we've missed. But also, and with bitter irony, the lethal

---

* Another record bowhead whales hold, according to the **Guinness Book of Records,** is the world's largest mouth.

effectiveness of the whaling industry in the nineteenth and twentieth centuries thinned out the population to such an extent that we've not yet waited
the 200 years we'd need to have a large population
of over-200-year-old whales yet.

Comparing mice and whales also demonstrates
one of the most famous observations in aging
biology: the bigger an animal is, the longer it tends
to live. Though there are many reasons why being
big might promote long life (or indeed necessitate
it, because growing big takes time), a significant,
simple factor is that if you're large, you're harder to
kill and eat.

Species which don't conform to this correlation
actually serve to corroborate the relationship between aging and extrinsic mortality. To keep it as
fair as possible, let's stick with mammals of a similar size. The house mouse, **Mus musculus,** weighs
about 20 grams, while the mouse-eared bat, **Myotis
myotis,** shares not just its ear shape but also its approximate weight with its namesake, with adults
weighing just under 30 grams.

The similarities do not extend to their longevity,
though: where a mouse might make it to three or
four years in captivity, the longest-lived mouse-eared
bat on record had just turned 37 when it died—and
that was in the wild, not coddled in a cage in the
lab. What lies behind this large difference in lifespan? Well, mice can't fly. It's not the pure joy of aer-
ial living which keeps the bats going for longer but,

rather, that being in the air keeps them safe from predators. There are far fewer threats up there, meaning that extrinsic mortality is sharply lower for a bat than for a mouse—which means, over evolutionary time, mutations have de-accumulated, antagonistically pleiotropic genes been selected against, and the advantages of disposing rapidly of the soma eroded. Today, bats live radically longer than mice in spite of being fairly close relatives, biologically speaking.

Another animal with a remarkable lifespan for its diminutive size is the naked mole-rat. Naked mole-rats are weird creatures: they look something like a penis with teeth and live in underground tunnels as "eusocial" colonies with a single breeding queen—more like ants and bees than mammals. At 35 grams, they are a little heavier than mice or mouse-eared bats, but they too can live over 30 years. They're also almost cancer-proof, in stark contrast to mice, and resistant to neurodegenerative disease. Their strategy of scurrying around underground is less romantic than taking wing, and has left them with tiny, beady eyes (it being too dark in their burrows for vision to be much use) and baggy, wrinkled skin (for ease of squeezing through tiny passages past other naked mole-rats—and which also, ironically, makes them look old even as youngsters)—but, nonetheless, it's worked. There are far fewer predators beneath the earth than roaming it, leaving the ancestors of naked mole-rats able to continually extend their longevity, too.

Humans, incidentally, are also very long-lived compared to other animals of similar size. Our secret to reducing extrinsic mortality isn't flying or burrowing, but probably relates to our large brains. These allow us to band together in complex social groups, share knowledge, build shelters, make tools, and so on, reducing the risk from external causes of death. As a result, we have evolved longer lives than our close relatives like chimpanzees—the verified chimp longevity champion, a female called Gamma, died at the age of 59.

So biologists can rest easy: in spite of superficially sounding paradoxical, the fact that animals live in risky environments is enough to relax the iron grip of evolutionary optimization in late life, causing aging to evolve. There's only one slight problem: a naïve reading of these theories predicts that all species should age. So how do animals which are negligibly senescent, like the Galápagos tortoise, fit into this picture? We've come full circle: now that evolution and aging are compatible, how can there be animals which **don't** age?

The theories we've discussed so far are incredibly useful, but they're inevitably simplifications of what happens in nature. If their assumptions don't hold, or if other factors we've not even considered come into play, different evolutionary strategies can result in unexpected aging trajectories.

Let's start with fish. Though they're scaly and live underwater, fish aren't such distant relatives of ours—they're still animals which, like us, have back-bones. However, unlike mice, whales or humans, female fish get bigger, stronger and far more fertile as they age. The fact that bigger fish are safer from predators than smaller ones means that their risk of extrinsic mortality isn't constant—it gets lower with age. They can also produce more or better eggs as they age, in some cases by ridiculous factors, with older fish producing dozens of times more eggs as young ones. These underwater matriarchs are known as BOFFFFs—big, old, fat, fertile female fish—and, in many species, are critical to fish populations. Fisheries are often sustained not by every young fish laying a few eggs, but by a handful of BOFFFFs churning out children at a rate of knots.

This reproductive strategy upends the assumptions which allowed aging to evolve in our thought experiments: increased survival and fertility in older fish compound to give BOFFFFs a grossly disproportionate opportunity to pass on their genes, creating a significant evolutionary incentive to keep them alive—the force of natural selection effectively reaches far further into adulthood. Perhaps the cold calculus of evolution will find that it's worth taking care of fish somas after all, and accumulated mutations or pleiotropic compromises which would strike down a BOFFFF are no longer acceptable to natural selection. Thus, fish could evolve whose overall risk of death doesn't rise with age—in other words, negligible senescence.

There are indeed some fish species which seem to be strong contenders for this. Of those in the running, the longevity crown goes to the rougheye rockfish, a pinky-orange Pacific seafloor dweller which can grow to be a meter long, weigh six kilos, and live to 205 years old, with chances of death that don't change detectably after reaching maturity.

Unfortunately for BOFFFFs, both commercial and recreational fishing prize large specimens. That means that overfishing hits the BOFFFFs especially hard and could therefore have a range of tragic consequences. First, there's the risk of fisheries collapsing, causing destruction to ripple through complex ecosystems connected to them. But it would also be a tragedy if species are eradicated before we have a chance to study them—not least to understand their unusual take on aging. And, even if we stop short of total destruction, preferentially catching BOFFFFs is causing some very unnatural selection to happen in these fish populations. Removing older breeding females will incentivize earlier reproduction, which could lead to genetic changes which introduce aging into these species.

As we've seen, some tortoises are negligibly senescent, too. The best studied are not from the Galápagos, but from Michigan. In a field study which started in the 1950s, scientists followed two types of turtle, Blanding's turtles and painted turtles. Hundreds of them have been marked and recaptured over decades, and no increase of death rate with time was observed in either species. When the study was shut down in 2007, the oldest fertile females were a couple of Blanding's turtles who were over 70 years old, with no external signs of encroaching frailty. The rationale behind the lack of turtle and tortoise senescence is probably similar to that in fish: older females are pretty safe from external threats (not least thanks to their protective shells), and highly fertile. Once again, natural selection has every reason to keep them alive, and the result is that they don't seem to age.

There are also weirder creatures, far further removed from humans than fish or tortoises, which sidestep senescence by other means. Hydra are a type of small, freshwater organism, made up of a centimeter-long tube with a sticky "foot" at one end, and a "mouth" at the other, encircled by flailing tentacles which grab their tiny aquatic prey and paralyse them with neurotoxic spines. They were initially of interest to science because of their startling regenerative capacity—chop off basically any bit of a hydra and a complete new hydra will grow from it. Only after this was it noticed that they survive an incredibly long time in the lab—to the point where, so far, hydra have outlived attempts to probe the limits of their longevity. They also show no signs of declining fertility or increasing risk of death no matter how long we keep them—and, based on the

death rates observed in lab-grown hydra, it's estimated that 10 percent of them would make it to
1,000 years old.

The regenerative capacity and off-the-chart lifespan of these tiny critters may not be unrelated.
Hydra violate the central assumption of disposable
soma theory—since any part of their body can go
on to make a new hydra, there is no distinction between body cells and reproductive cells. In effect,
they're all reproductive cells, so evolution doesn't
consider any of them disposable. This is a hack which
is only going to work with very simple forms of
life—complex life from insects to humans all undergoes a one-way conversion of reproductive cells to
body cells, allowing us to have such diverse tissues
and organs—but it shows how almost no assumption is safe in the face of real-life biology. Nature
will continue to outwit our theories for some time
to come—and aging itself, if it has to.

There are also evolutionary pressures which could
select for longevity almost directly, not as a side effect of high late-life reproductive capacity or blur-
ring the lines between soma and reproductive cells.
Enter what's thought to be the longest-lived multicellular life form on Earth: a bristlecone pine at a
top-secret location in California's White Mountains.
A core taken from this tree's trunk in the late 1950s
had nearly 5,000 rings. The tree is still going strong
today, at an estimated 4,850 years old. That means it
germinated at the start of the third millennium BCE,

when Stonehenge was just a ditch and a few small stones, and work on the pyramids was yet to begin.

How a tree could evolve to be able to outlive whole civilizations isn't entirely certain, but one theory is that it has to do with competition for space. Bristlecone pines live in arid, exposed environments where all the habitable locations are already occupied by incumbent adult trees, meaning opportunities for saplings are rare. Basically, you need the tree next door to die in order to free up a spot where your descendants could set up shop. Thus, the only way to pass on your genes is to outlive your neighbors, starting an evolutionary arms race whose endpoint is extreme longevity. Obviously this kind of logic doesn't apply to animals which can simply walk to another location if things are getting crowded— but it is another example of how a simple quirk of natural environment can significantly affect the evolution of aging.

Depending on the relative strength of all of these factors and more, negligible senescence doesn't seem so bizarre an outcome. Change the relative likelihood of survival and importance of reproduction by organisms of different ages, and evolution will custom-build you a life course which optimizes for that—with a huge spectrum of different results, from those mayflies which live for only a few minutes, to trees which survive for thousands of years.

If it makes evolutionary sense in some cases to have a risk of death which is constant with age, could we take the next logical step—is there the possibility of **negative** senescence, a risk of death which **decreases** with age? Though we don't know of many life forms lucky enough, there do seem to be some which possess it: for example, the best data we have on the desert tortoise suggest slightly negative senescence throughout its adult life. There's probably nothing special about negligible senescence, and it would be slightly weird if there was a hard floor of zero for change in risk of death with age. There are probably more creatures with negative senescence out there, just waiting for careful demographic studies to uncover them—assuming human consumption or environmental destruction doesn't wipe them out before we get the chance.

Thus, the evolutionary theories of aging don't just explain why some animals get old, but also open the door to slowing or even removing senescence entirely. There are real-world examples of organisms which sidestep aging, and we have solid theories about the forces which have subdued their tendency to degrade with time. For anyone interested in changing the course of human aging, this is incredibly exciting news: negligible (or even negative) senescence not only doesn't break the laws of physics—it doesn't break any laws of biology either.

Nature also shows us that lifespans vary by a large amount between even quite closely related species. Comparing mice to bats and naked mole-rats is a striking example of how animals with similar size and relatively recent common ancestors can nonetheless age in very different ways. This shows us that aging isn't some immutable, inevitable process: these variations between animals prove that learning to evade aging is possible. They also provide inspiration: comparing the biology of species with different aging rates will allow us to identify the genes and mechanisms which promote longevity in long-lived species, and to try to develop drugs or treatments to mimic them.

However, the most important thing that an evolutionary explanation provides is insight into what aging is and what it isn't. We now know that we don't all have some ticking, internal clock, programmed to kill parents to make space for their children. It might be easier if we did—all we'd need to do is find the time bomb in our genes, defuse it, and aging would be cured.

Instead, aging is an evolutionary oversight: a result of mutations accumulated which worsen fitness in old age but evolution can't get rid of; antagonistically pleiotropic genes that maximize reproductive success in youth even if they have unfortunate unintended consequences in later life; and mechanisms that prioritize having children over maintaining our disposable somas. There is therefore no reason to expect that aging should have a single cause— indeed, we should expect it to be made up of a collection of synchronized but only somewhat related processes. It's our job to identify these processes, and treat them.

However, this kind of can-do attitude has only been tenable for the last couple of decades. The evolutionary theories of aging were developed in the middle of the twentieth century and, in spite of being significant strides in our understanding, they had an ironic and unfortunate side effect. Aging had long been largely ignored by biologists, seen as a phenomenon of gradual deterioration not really amenable to study. These evolutionary theories underlined this hopelessness: they suggest that many processes are likely to contribute to aging, with no obvious limit on the number of contributors. There could be hundreds or even thousands of different factors, interacting in myriad different ways, all conspiring to end us. Evolutionary theory doubles down on the idea of aging as a process so tangled and multifaceted that it's unlikely ever to be understood, let alone treated.

If we are to be confident about understanding aging and, ultimately, curing it, we need to be convinced that it can be tackled on something other than evolutionary timescales. The discoveries which permit us to imagine doing so are the subject of the next chapter.

3

The Birth of Biogerontology

Modern aging research is often referred to as biogerontology—the biological subset of gerontology, which covers everything from medical care of the elderly to the social aspects of growing old. It's foolhardy to pick a precise start date for a scientific field, but the emergence of biogerontology as a distinct, sizeable discipline arguably began in the 1990s—shockingly recent for a field which concerns itself with one of the most significant, near-universal phenomena which afflict living things.

It's hard to pinpoint exactly why aging research was a backwater of biology for so long. Pre-existing skepticism that aging was just too complicated for serious study, underscored by an evolutionary understanding of aging which implied that an almost infinite number of processes contributed to it, certainly played a part. Then there are socio-scientific factors: no scientists (or politicians funding them) have parents or grandparents who died of "aging" per se, meaning that research into diseases like cancer which are directly responsible for death tend to get more attention. Scientists also tend to cluster on research topics: there are trends and fads in science just as there are in music and fashion. Perhaps aging research had a low profile partly because, for whatever reason, it never gained scientific critical mass?

Looking back at its history, one tempting narrative is that scientists needed at least some proof before they were willing to start work on aging: proof that aging could be altered, and proof that it could be done in ways which were scientifically interesting, and tractable in the lab. Two sets of experiments stand out as having provided this proof, serving as the foundations of modern aging biology. This chapter is thus split into two parts: we'll start with the long-lived hungry rats which provided the first direct evidence that aging can be altered, and then turn to long-lived, genetically modified worms which showed that not only could it be modified, but that you could do so in remarkably simple ways—in fact, by changing a single letter of DNA.

LIVE, FAST, DIE OLD

Food is delicious.

It's not surprising. For billions of years, since our most distant evolutionary ancestors, all life has been locked in a struggle to acquire enough food to live and reproduce, or perish. Genes which equip a creature with a desire to seek out and consume food confer a huge survival advantage. Consequently, our brains are wired to enjoy eating, and be wildly distracted by hunger until our need for food is sated. However, evolution hasn't been so careful when it comes to placing an upper limit on how much we should eat. When you don't know how long it will be until your next meal, it makes sense to gorge yourself should the opportunity arise.

In the early twentieth century, humans were finally starting to move beyond this natural state of bare subsistence in large numbers and, now that people finally had a choice about what and how much they ate, scientists were beginning to take an interest in the effects of nutrition on health. It was, quite unexpectedly, from this emerging field that the first reliable results in aging biology would appear.

Scientists experimenting with the effects of nutrition on growth noticed that underfed animals attained smaller overall size—so far, so obvious— but they also seemed to live longer. These early results were suggestive but not conclusive—the number of animals in each experiment was small, and their diets weren't carefully controlled for calories, protein, vitamins and minerals. However, their findings were enough to intrigue American scientist Clive McCay, an assistant professor of animal husbandry at Cornell, who set about performing the first meticulous experiment which was also large enough to give convincing results.

McCay took 106 rats and split them into three groups, with one group eating what they liked, one on a restricted diet which started immediately after weaning, and another which enjoyed two weeks of free-for-all before having their rations cut. Critically, and in contrast to previous work, McCay kept everything as constant as possible to make sure that the dieting rats got all the vitamins and minerals they needed—the only difference in their food intake was number of calories.

The study set records for rat longevity. On a normal diet, the longest-lived male rat survived 927 days. Many of his compatriots on restricted diets were still going: the last one standing died at 1,321 days old—a 40 percent extension of maximum lifespan. The average lifespan of male dietary restricted rats very nearly doubled, from 483 in the well-fed group to 894 days.*

Not only were they living longer, the dietarily

* The data from the female rats in the experiment are somewhat more confusing—not least because some of them died very early in the experiment during a particularly hot spell, distorting the results.

restricted rats were healthier, too. When they died, the rats were given post-mortems and, upon dissection, those on lean rations had much better-looking lungs and kidneys. It was already known that restricting food intake could reduce rates of cancer in rats and mice, and McCay's findings doubled down on this: none of the restricted rats developed tumors until they were put on a normal diet near the end of the experiment. And, more evocatively, the rats just **looked** healthier: "the hair of the animals retarded in growth remained fine and silky for many months after that of the rapidly growing animals had become coarse," he wrote in a 1934 paper. These results showed very clearly something which had been dreamed of for thousands of years of human history: the aging process could be slowed.

It is genuinely shocking to modern eyes that news of these findings didn't reverberate around the world and lead to widespread, far deeper investigation of this phenomenon—for the first time in history, the process of aging had been slowed down! Unfortunately, for whatever sociological or scientific reasons, it didn't. It might have been that aging wasn't a major concern in the 1930s, while growth and development were. Life expectancies in the United States at the time had only just reached 60 years, and the specter of infant mortality loomed large in recent memory. The focus was on ensuring a healthy childhood rather than a healthy old age: McCay's 1935 paper gives as much prominence to effects on rat growth and development as it does to their lifespan. Over the next few decades, a patchwork of studies did carry on investigating the link between diet, health and longevity, not least by McCay himself, but it was 50 years before dietary restriction—or DR—finally began to be properly investigated.*

These subsequent studies showed that this phenomenon, far from being a quirk of rat physiology, is one of the most universal in biology. The number of species in which dietary restriction works is incredible: we've tried it successfully in yeast, the microscopic, single-celled fungi used in baking bread and brewing beer; tiny nematode worms; flies, spiders and grasshoppers; guppies and trout; mice, rats, hamsters, dogs and maybe rhesus monkeys. (Why only maybe? We'll come to that.) Some of the techniques used to restrict diets in other organisms are unusual. Particularly with the smaller creatures, you have to get inventive when cutting their rations. Nematode worms slurp up bacteria

---

* You'll often see dietary restriction referred to as "calorie restriction" or "caloric restriction," abbreviated as CR. Given that, starting with McCay's experiments, the importance of optimal nutrition (ON) as well as cutting back calories was recognized, it's sometimes known as CRON, and its disciples affectionately referred to as CRONies. Nonetheless, I'm going to call it "dietary restriction," or DR, a bit of pedantry needed because modern research has called into question whether it's the calories themselves which are important, or other aspects of diet like protein, or individual amino acids. We'll return to this in Chapter 10.

as they slither around, so you have to thin out the "bacterial lawn" on which they graze, and add just enough antibiotics to stop the bacteria from multiplying and turning famine to feast. My favorite methodology was used on water fleas: it involved using pond water to dilute the delicious "manure infusion media" on which they usually dine, and extended their lifespans by a cool 69 percent.

The incredible universality of this effect, from single cells to complex mammals, is an example of "evolutionary conservation." It implies that this response to reduced food levels is ancient, a piece of biology so fundamental that it has been kept in every kind of creature even as the tree of life blossomed into endless forms most beautiful. And its implications are tantalizing—if everything from water fleas in dilute manure to dogs on rationed meals live longer and in better health, could it work for humans, too?

There is a catch: despite its evolutionary conservation, the size of the DR effect varies wildly in different organisms. Single-celled yeast can have its five-day lifespan increased by 300 percent; nematode worm **C. elegans** lives 85 percent longer under DR; fruit flies 66 percent longer; mice 65 percent longer; mouse lemurs (which are primates, like humans, but are rather distant relatives, and weigh just 50 grams) can have their six-year lifespans extended by 50 percent; rats come in at 85 percent or so (as we've seen); while the best effort in dogs achieved just a 16 percent lifespan increase. Cost and practicality mean that we've not tried many large, long-lived animals, which makes extrapolating any trend hidden in these statistics to humans (which are, in the scheme of things, large and long-lived) quite tricky.

This debate might have been resolved by the recent conclusion of two studies in some close evolutionary cousins of ours: rhesus macaques, a species of monkey with a maximum lifespan of around 40 years. The good news is that DR did seem to increase healthspan in both studies; the bad is that the effects on lifespan were more ambiguous, and certainly rule out results as impressive as those in worms, rats or lemurs. Human studies have been too short to give any definitive answers on lifespan or healthspan, though short-term markers of health like blood pressure, cholesterol levels and markers of inflammation do seem to be improved.

We'll return to the rhesus macaques, and whether we should all be practicing DR, in Chapter 10. However, for now, suffice to say that it's no slam dunk: if there is a trend, it probably shows that the effect weakens as organisms approach humans in size, longevity and complexity. Quite apart from any advanced biological argument, given the diversity of diets around the world we'd surely have noticed by now if DR doubled human lifespan—there would be some ascetic religious sect living twice as long as the rest of us, and even modest differences in diet would have much larger impacts on health and longevity than we actually observe.

However, while the dietary debate rumbles on for monkeys and humans, DR is hugely significant in the history of biogerontology. Its foundational contribution, through experiments by McCay and others, was to show that aging can be slowed—an unambiguous demonstration of a critical fact without which it would be very hard to convince skeptical scientists that research into aging is worth pursuing. In more recent times, it's also been pivotal in beginning to decode how aging works under the biological hood.

A renewed interest in DR research in the last few decades, together with expanded tools for molecular biology, allowed us to examine what happens when food is scarce. The results of these studies actually provide reason to hope that the response to DR might be truly universal: the molecular machinery which implements it is shared in every species we've looked in, from yeast to people. When any of these organisms eats something, an almost identical system of molecular detectors and signals alerts the cells to the incoming nutrients and sets about making use of them—storing some for later use, putting others immediately to work building new cellular components, and so on. In the absence of nutrients, this system sends these processes into reverse, telling cells to rein in their manufacturing and hunker down while raw materials are less abundant.

Why might this response to low food levels be so carefully evolutionarily conserved? The most popular idea is based on disposable soma theory, and centers on how animals weigh the competing energy requirements of somatic maintenance and reproduction. If you find yourself on a restricted diet, and can only choose one, maintaining your body is the obvious option: rather than blowing your calorie budget reproducing one final time in desperation, you can live on to reproduce another day; and it also means that your newborn children won't arrive into the world during a famine and promptly die themselves. Thus, evolution has selected for animals who allocate more resources to carefully maintaining their bodies when times are tight, slowing the gradual falling apart that is the aging process. When food is plentiful again, reproduction takes priority and aging returns to its original pace.

We'll meet some of the molecular protagonists later in the book—they range from insulin, which you may have heard of as the hormone which maintains blood sugar levels and whose manufacture or detection goes wrong in diabetes, to more esoteric actors you might not have, like mTOR. Having a treatment that reliably slows the rate of aging has also helped illuminate the processes of aging, by pointing to biological changes which are slowed under its influence. Were it not for countless hungry mice, flies and worms in labs around the world, we would know far less about aging today than we currently do.

Whatever else DR experiments show, they tell us loud and clear that aging is not some inevitable, immutable, unstoppable process. How quickly animals age can be varied by this (perhaps deceptively) simple intervention. Why this didn't result in a biogerontological revolution before now is academic: DR, along with the many other potential treatments we'll talk about in this book, provides incontrovertible proof that aging can be manipulated. We should celebrate dietary restriction for providing us with the poster rats of anti-aging medicine.

One problem remained: DR showed us that aging can be manipulated without the need to wait around for your descendants to evolve into some different, slower-aging species—but it was still not easy to decipher. Aging still suffered from its image as an inscrutably complex process of wearing out. The fact you can cause animals to wear out more slowly doesn't do much to reduce that complexity, or suggest that its treatment is any more plausible. Dietary restriction played a critical part in the gestation of biogerontology, but it would take another breakthrough before it could be born.

THE WORM HAS TURNED 150

One of the most important stories in aging biology starts in a rather inauspicious location: a compost heap in Bristol, UK, in 1951. Wiggling through the dirt was a population of nematode worms, which are arguably biogerontology's most historically significant organism, responsible for transforming the field into a serious scientific discipline. Without those West Country worms, we could well be decades behind where we are today.

A decade after their composty beginnings, biologist and later Nobel laureate Sydney Brenner was looking for an animal in which to study neural development which was simple enough to have a hope of understanding it. His first experiments were done with a type of nematode which he found in the soil in his back garden in Cambridge and he named them N1—nematode 1. However, he was eager to find the best worm for the job, and wanted to audition other candidates before proceeding. The Bristolian worms were the eventual winners, and they were christened N2, which is less of a mouthful than their full biological name, **Caenorhabditis elegans** (or **C. elegans** for short). These nematodes, tiny, millimeter-long worms, transparent, unassuming, and barely visible to the naked eye, are now one of the most successful "model organisms" on the planet.

Model organisms are one of the key tools in

modern biology. These are creatures which are used as testing grounds for everything from drugs to blue-skies biological theories. The idea of using a model organism is to simplify a problem, both conceptually and experimentally, and allow us to glean insights which can then be used in more difficult, complex organisms like humans. The classic quartet in aging biology (and many other fields, too) is yeast, worms, fruit flies and mice, in ascending order of biological similarity to us.

The key difference between these nematode worms and mice or people is the obvious one: it's a vastly smaller animal. Comprising just under a thousand cells rather than trillions, their bodies are small enough that we can get a handle on the behavior of every single cell. There's even a project called OpenWorm trying to build a full cell-level computer simulation of **C. elegans**—something we can only dream of doing for humans for now.

The experimental advantages are also significant. **C. elegans** comes to the rescue when human studies would be too inconvenient, too lengthy or just an ethical nightmare. These worms grow up, reproduce and die in just a couple of weeks, massively speeding up experiments, you can grow dozens of them in identical conditions in a small dish in the lab—a setup which humans tend to object to—and we also have fewer qualms with worms when it comes to genetically modifying them just to see what happens.

The first worm experiments sound haphazard

and primitive compared to modern science, with its precision gene-editing and sequencing. The old technique was to take some N2 worms (N2 is still used to refer to the "standard" strain, the HB pencil of **C. elegans** varieties) and expose them to a nasty chemical which induces random mutations in their DNA; take the thousands of mutated eggs they produce and grow them each individually into adults; breed dozens of identical copies from each one; and finally, check to see if any of your randomly mutated worms do anything interesting—in this case, watch them for a few weeks to see how long they live. If one of these mutants lives longer than normal, then whatever changes lie in their DNA might help us to understand the genetic basis of longevity.

In 1983, scientist Michael Klass was starting to lose faith in this protocol, after several years testing a staggering **eight thousand** strains for longevity mutants. He had discovered just eight which lived longer than normal, and found reasons to dismiss all of them as uninteresting: two spontaneously went into a worm-specific form of suspended animation called the dauer* state, which is probably cheating (even if humans **could** do something similar, living longer by spending decades in a strange environmentally hardened cuticle probably isn't what most people have in mind); one had a defect which

---

* "Dauer" is a German word which literally translates as "duration," but in this context means an enduring or permanent state.

seemed to stop it from sensing and moving toward food; and the other five all appeared lethargic under the microscope. Klass suspected that these latter six strains, through either a sense of smell failure or general lethargy, were eating less than their N2 counterparts. It was by now common knowledge that eating less makes animals live longer—so all he'd done was rediscover dietary restriction via an incredibly laborious and roundabout genetic route.

Klass's failure to identify a longevity mutant fitted right in with the prejudices of the time: aging, as we've seen, is down to many different genes which have terrible effects in late life and accumulate in our DNA either by chance or because they conferred an advantage in youth. It was thought that there must be dozens or even hundreds of such genes, each whittling away at an old organism's life chances—if one single longevity gene could make any difference, why wouldn't evolution just turn it up to eleven and marvel at its long-lived creations? And wouldn't we expect to see occasional mutant people who live far longer than the rest of us?

Klass's results seemed to confirm this thinking: mutating a few genes in a worm couldn't extend its life, other than by the back-door route of depriving it of food. He ended up quitting academic science in frustration, but his colleague Tom Johnson took up the quest with remarkable tenacity. Johnson was hoping that the worms' life extension was real, and that he could use it to double down on the existing dogma and prove that aging was controlled by many different genes. He knew that these mutation-inducing chemicals typically introduced about twenty errors in each worm's DNA, so it was quite possible that the long-lived worms possessed a string of genetic alterations, some positive, some negative, and all ripe for investigation.

Step one was to work out if the worms' impaired feeding was important. He started by breeding the mutants with N2 worms, the first step of a painstaking process to isolate the genes responsible in the days before genome sequencing. He managed to produce some worms which ate as much as normal, but still had long lifespans. DR ruled out, he bred some of these long-lived and well-fed worms with N2s. To his astonishment, the worm-children produced by this union had normal lifespans.

The simplest explanation for this observation was that a single gene **was** responsible for the life-extension observed.* If there were many genes involved, it was highly unlikely that the entire effect would vanish in the first generation—you'd expect the lifespan of the cross-bred worms to be somewhere in between N2 and the long-lived mutants. Then breeding the long-lived mutants with each other didn't enhance lifespan further—suggesting

---

* It also requires that the mutation is "recessive"—meaning that it needs to be present in two copies, one from each worm-parent, to cause the lifespan extension.

that they all shared either the same or a very similar genetic mutation.

Eventually, Johnson convinced himself that it was indeed a single gene responsible for the long lives of these worms. He published his findings in 1988, naming the gene **age-1**. Its effects were impressive— the worms' lifespans were increased by 50 percent, from two weeks to three. That's equivalent to finding a single mutation which means humans could habitually live to 120 rather than 80.

Unfortunately, he failed singularly to convince the rest of the biology community. Many biologists thought the work could be erroneous or, if not, just a weird quirk of nematodes with limited relevance to other species. Even were it true, there was cause to doubt its deeper significance—the **age-1** mutants were not only long-lived, but also had substantially decreased fertility. Far from casting doubt on the evolutionary theories of aging, Johnson had just confirmed them with a perfect example of disposable soma theory—a single gene which extended life, but by redirecting resources from reproduction to somatic maintenance.

Though it didn't cause fireworks at first, discovering **age-1** did light a fuse. It inspired another worm biologist, Cynthia Kenyon, to go out searching for more longevity genes. In 1993, she found another longevity mutation, this time in a gene called **daf-2** which was already well known to worm biologists—it was discovered in another of those random mutation experiments, and worms with **daf-2** mutations were especially eager to enter the long-lived dauer state. Kenyon's experiments showed that, if raised at a cool temperature to stop them from becoming dauers, these worms lived longer than normal worms: whatever mechanisms allowed dauers to tough it out for months awaiting more favorable conditions work in adult worms, too, extending their lifespan. And the results were spectacular: **daf-2** mutants lived twice as long as normal worms.

Further work on **age-1** and **daf-2** mutants showed that they really were delaying the aging process. While two-week-old N2s spend their final days looking haggard and barely moving, their longevity mutant contemporaries look youthful and fresh, and slither around rapidly. Late-life decrepitude— which is so severe as to be visible to the untrained eye through a microscope, even in these unfamiliar creatures—didn't kick in until shortly before their own demise, a couple of weeks later. These mutations don't just extend life—they slow down the aging process itself.

Where **age-1** could be dismissed as a quirk of worm biology, a second gene with a plausible mechanism of action and an even more impressive boost to worm lifespan did a lot to dispel these doubts. The scientific importance of this discovery is obvious: DR experiments had already shown that aging could be manipulated, but to alter it by changing a single gene is astounding. How could just one gene have such a dramatic effect, seemingly across the whole spectrum of age-related changes?

But perhaps the bigger effect was cultural. This finding opened up the study of aging to the precision techniques of modern genetics and molecular biology. No longer was aging a process so messy that it couldn't be studied—if you can control it with pinpoint alterations to single genes, then the process was suddenly open to the kind of methodical tinkering which might allow scientists to decode it. This discovery was a landmark, showing that aging was not just malleable, but also intelligible. Aging research, formerly viewed as a scientific dead end, now took to the limelight. The modern scientific investigation of aging was born.

The story of **age-1** and **daf-2** doesn't end there. A gold rush of worm genetics uncovered many more mutants which affected aging. The record for longest-lived worms was repeatedly smashed by worms carrying different mutations in different genes. With poetic symmetry, the current reigning champion is **age-1**—the same gene, but a different mutation to the original eighties Klassics. Worms carrying it live for an average of 150 days—a jaw-dropping tenfold life extension over N2 worms. In the end, the confirmatory experiment lasted almost nine months, with the final **age-1 (mg44)** worm dying after 270 days. While arguably a slightly facile comparison, that's roughly equivalent to a human living 1,500 years.

And because this experiment was performed in the

mid-2000s—well into the era of DNA sequencing—
we now know something even more amazing about
age-1 (mg44). The mutation which gives rise to this
incredible longevity results from a change in a single DNA letter: 1,161 bases into the **age-1** gene, an
A replaces the usual G. That turns a TGG sequence
into TGA which, in the language of DNA, means
"you're done, stop reading." As a result, the AGE-1
protein* is about a third of its usual size and missing
crucial components. So useless is this truncated protein that, like the first third of a car, with one and a
bit wheels and a few random parts of the engine, it
might as well not be there at all. Previous **age-1** mutants merely made the protein less efficient at its job,
and consequently had less spectacular effects—but
its complete absence dramatically extends lifespan.

What ghastly poison is AGE-1 that its presence
cuts worm lifespan by a factor of ten? And why on
earth do worms produce this deadly stuff inside
their cells? Cynthia Kenyon refers to **daf-2** as "the
grim reaper"—which makes **age-1** the Terminator
crossed with Genghis Khan.

It turns out that both **age-1** and **daf-2** are part
of the machinery which allows worms to respond
to changes in food levels in their environment: this
is a crucial part of the system which mediates the

---

* Pedantic typography alert: the **age-1** gene provides the DNA instructions to build the AGE-1 protein. The nomenclature varies from
species to species (of course it does), but worm gene names are usually lowercase italic, versus capitalized normal text for their products.

evolved response to dietary restriction. DAF-2 is an insulin receptor, a molecule which sticks out of the surface of a cell, looking out for insulin to grab on to. Insulin, remember, is the hormone responsible for controlling blood sugar levels in humans, and telling the cells of the body to use or store the nutrients coursing through our bloodstream after we've had a meal. A family of 40 insulin-like molecules does basically the same job in worms, telling cells to change their behavior when there are nutrients around to be used.

If the DAF-2 receptor detects insulin, that tells it that food is plentiful, and it can set in motion processes like growth and reproduction to propagate the species. If it doesn't detect insulin, then times are lean: if you're a young worm, it might be worth taking time out as a dauer; in an adult worm, it fires up processes to maintain the worms' bodies, and hopefully outlast the famine. The DAF-2 receptor detects insulin, and then the AGE-1 protein spreads the good news and gets the processes of rapid reproduction (and rapid aging) going. If you imagine that DAF-2 is the accelerator pedal which insulin can push to speed up growth, reproduction and aging, then AGE-1 connects the pedal to the throttle which puts fuel in the engine. Remove the pedal or the connector, and there's no way for insulin to stamp on the gas, and aging is slowed down whether you have either mutation—or both.

The end result of this genetic alteration is thus

that the worms' cells end up behaving as though
there's a famine, when actually food could be plentiful. So, in a sense, Klass was right—these genetic
alterations **were** DR by the back door, conferring
many of the benefits of eating substantially less that
we explored earlier in the chapter. The difference is,
this is a fascinating, molecular backdoor which gives
insight into how aging works at the cellular level,
not a clunky, roundabout way to actually reduce the
worms' food consumption.

Worms deserve their place in history for firing up
the science community about aging, but you'd be
forgiven for not getting overexcited about the relevance of these Methuselah worms for human medicine. However, there is reason to keep an eye on
findings in model organisms: evolutionary conservation. While obviously yeast, worms, flies and mice
differ from us in many, many ways, they share an
awful lot of fundamental biology with each other—
and with us.

The genes responsible for these incredibly long-
lived worms are one such common feature. Mutations
in the insulin signaling pathway and growth hormones are also found in long-lived strains of yeast,
fruit flies and mice. These include the Laron mouse,
which has a mutation in its growth hormone receptor gene—and the longest-lived of which died just a
week before its fifth birthday. Because this mutation
affects growth hormone, the mice involved mature
more slowly and end up much smaller than mice

without the mutation, but they go on to live longer and in better health.

In fact, the Laron mice were genetically modified to mimic a condition discovered in humans, Laron syndrome. Found primarily in people living in remote villages in Ecuador, this genetic mutation means that the villagers are very small—typically a meter or so in height—but it also seems to keep them almost completely free of cancer and diabetes. Unfortunately, it's very hard to work out whether this confers the longevity benefit that the worms and mice enjoy, and that their own freedom from cancer and diabetes suggests: a study found that the life expectancy of those with Laron syndrome is pretty much normal, but 70 percent of the deaths in the group were from non-age-related causes, including 13 percent due to alcohol and 20 percent from accidents. It's unclear if their lives would be longer in the absence of these significant drags on life expectancy.

These mutations in insulin signaling and growth hormone genes are a bit like genetic versions of dietary restriction, but sidestepping the need to actually restrict diets: they trick cells into thinking the cupboard is bare when, actually, it may not be. So, while it's an evocative description, these genes aren't really the grim reaper or Genghis Khan—they're a vital survival mechanism which allows worms, mice and humans to alter their metabolism in response to changing conditions in the wild.

We know how vital thanks to more worm experiments: if you put mutant worms into competition with wild ones, you rapidly find out why the grim reaper gene is necessary. On a plate of both N2 worms and **age-1** mutants where food levels were varied to simulate the feast-and-famine conditions you might find in the natural habitat of **C. elegans,** the N2s (with their grim reaper intact) rapidly outcompete their mutant cohabitants. A similar experiment pitting **daf-2** mutants against N2s conducted in soil, rather than the usual barren environment of an agar plate in the lab, showed that the nonmutant worms actually lived longer in real-world conditions. Evolution, as ever, is all about trade-offs: in this case, the naturally occurring N2 worms accept a shorter life in paradise in exchange for a more reliable lifespan and better reproductive potential in the real world.

Cosseted in the lab, living free from competition on a plate of genetically identical worms, these longevity mutations bestow worms with lifespans which would be astonishing in nature. This is commonly used to suggest that some of the life- and health-extending interventions we'll discuss in this book are not practical out in the real world, because they require trade-offs which make creatures more fragile in subtle ways that don't show up in the lab. However, there's a much more optimistic take on this: for humans, in the rich world at least, hygiene, healthcare, a steady food supply and so on mean that our pampered lives far more closely resemble those of worms in a Petri dish, isolated from natural hazards, than they resemble those of wild animals, be they worms in the soil or prehistoric people. We effectively live in a giant lab environment of our own construction, and one for which our genes, refined by natural selection for the environment in which we evolved, are not necessarily optimized. This could mean that we may, like **C. elegans** on a lab bench, be able to benefit from substantial changes to our rate of aging.

Though it seems unlikely that specific genes discovered in worms will lead to any direct improvements in human longevity, their importance to the birth of biogerontology cannot be overstated. What was thought for decades to be an impossibly complicated process, out of the reach of lab biology, could be substantially altered by changing a single gene—indeed, a single letter of DNA. This placed aging firmly within the grasp of lab biology.

Mutating a gene in a model organism is one of biologists' favorite ways to make sense of a problem. You can think of it like modifying or entirely removing one component of an engine and seeing what happens. The consequences can start to tell you what that component is for, and how it affects the parts it's connected to, providing data with which you could ultimately uncover how the engine works. In a human-designed machine, this is a grossly inefficient route to understanding: the result will probably just be that it stops working, leaving you none the wiser as to the component's function. In biological systems, which are messy, interconnected and evolved with layers of redundancy which often makes them robust in the face of small alterations, the results of a small change can be far more surprising—such as a massive increase in lifespan.

If you can alter lifespan so dramatically with the change of a single gene, this gives us the ability to ask a huge range of new questions. What does this longevity gene do? What genes does it work with? If you mutate those genes, does the effect get bigger, smaller, or stop working altogether? By tugging on these threads, biologists were able to start investigating the processes which drive aging in a far more systematic way than when they didn't know where to begin. We now know of over 1,000 genes which can increase lifespan in various organisms—including 600 in **C. elegans.**

That's why these developments signaled the beginning of a new field. Aging was now something which could be intervened in, poked, prodded and studied. Studying it was no longer a weird pastime, largely ignored by mainstream biology, and career suicide to take an interest in. We could finally answer the age-old question of what aging is, not just in the general evolutionary sense of it being a collection of processes involved in deterioration, but in a nitty-gritty, cellular and molecular catalogue of what goes up, what goes down, what might be a cause or an effect. That's exciting from a scientific point of view, but it's also critical if we are to have any hope of treating it. Next, we'll turn to what this exciting new science uncovered.

4

Why We Age

Over the last century there have been dozens of theories of aging purporting to explain why we grow old and die, many of which have buckled under the weight of contrary evidence. Rate-of-living theory, the DNA damage theory of aging, mitochondrial free radical theory, garbage catastrophe theory: there was a running joke that there were more theories of aging than scientists to work on them. Which, given the size of the field in the past, might not have been so far from the truth.

One particularly delightful theory of aging is that all animals have a fixed number of heartbeats in our lifetimes. Mouse hearts beat an incredible 500 times per minute, while Galápagos tortoises' tickers tick almost 100 times slower at just six. Can it be coincidence that Galápagos tortoises live 175 years, almost 100 times longer than a mouse's two-year lifespan? If you collect data from a wide range of different species, there's a striking pattern—heartbeats per lifetime is remarkably constant, from rats and mice to elephants and whales. We each get about one billion beats, and then we expire.

The theory seems to work within species as well as between them: doctors know that patients with a higher resting heart rate are at increased risk of death. Having a resting heart rate of 100 beats per minute doubles your annual risk of death compared to those people whose hearts beat at 60 bpm. Could this be because they're burning through their allotted beats with unseemly haste?

While this idea is intriguing, it's probably of limited practical value. For a start, the relationship between different animals isn't as tight as the headlines suggest: I'm sure some of you did the math for the mouse and the tortoise, and it comes out closer to half a billion beats; and humans are massive positive outliers, with roughly three billion heartbeats per lifetime. It could also just be a coincidence—we've already seen that bigger animals live longer, and there's known to be a relationship between body size and heart rate, so maybe body size is the causal factor here. Finally, it's not clear how or indeed if this would translate into a treatment—though drugs exist to lower heart rate, this would arguably be treating a symptom of the diseases or lack of physical fitness that lead to high heart rate, rather than the cause. And there's obviously a limit to how low you can go—we could imagine using medicine to move a patient from 80 to 60 bpm, but there must come a point where the heart simply isn't beating fast enough to supply the body with blood. (The best treatment for a high resting heart rate, incidentally, is almost certainly to do more exercise.)

What biogerontologists, doctors and the rest of us really want to understand is the underlying causes of aging—the cellular and molecular changes which underlie the downstream consequences for organs like your heart. With the ability to intervene in aging both genetically and via diet, plus newfangled molecular biology to pick through the resulting changes, modern biogerontology has had the chance to study the aging process in far more detail than counting heartbeats. Over the last couple of decades, scientists have uncovered the changes which occur in our bodies as we age, and begun to piece together a coherent picture of how these are connected to the diseases and dysfunctions that accompany the process. This is not just for scientific interest, but also because things which are closer to the root causes are more amenable to treatment: slowing your heart rate to zero to stop you from aging doesn't make sense, but the idea that eliminating a root cause of aging might improve your health does.

This new understanding of the underlying causes
of aging has shown us that, as predicted by the evolutionary theories we studied in Chapter 2, aging
isn't one single thing—but nor is it thousands. We
now know enough to attempt to place aging-related
changes into categories. Most excitingly, there are
few enough that we can hope not only to explain
what drives the aging process, but potentially come
up with treatments to address it.

There have been several attempts to systematically classify theories of aging, but two modern ones
stand out because not only do they provide a classification system, but they do so explicitly to guide
devising treatments for aging. The first, originally
published in 2002 and boldly entitled "Strategies
for Engineered Negligible Senescence" (SENS for
short), was devised by maverick biogerontologist
Aubrey de Grey. In its current form, SENS identifies seven differences between old bodies and young
which de Grey suggests are the fundamental causes
of aging. It's fair to say that it was, and remains,
controversial. Because it was motivated specifically by treating aging, his "seven deadly things" are
types of age-related "damage" grouped because he
envisages a type of treatment to tackle each one. If
we could hit them all at once, he claims we could
postpone aging for long enough to buy us enough
time to develop the next iteration of SENS, and so
on—which is why he calls them strategies for negligible senescence. If we could manage this, de Grey

argues, we could look forward to thousand-year-plus lifespans—claims which have understandably raised a few eyebrows among scientists. Some of the proposed treatments were outlandish, and even the more plausible ones were speculative because at that time none existed, let alone had been shown to work—but the idea of grouping age-related changes like this is a good way to build a framework for treating them.

The second, published in 2013, is "The Hallmarks of Aging," and enumerates nine changes which fit three criteria. First, they need to increase with age: if they don't, how could they be causing aging? Second, accelerating a hallmark's progress should accelerate aging, and, third, slowing one should improve it—these two criteria are an attempt to separate things which are merely associated with aging from things that are actually contributing to it. Finally, these hallmarks also come with suggested interventions which could slow or reverse their progression, thus slowing or reversing that aspect of aging, and hopefully putting the brakes on the process overall.

These two classifications have a lot in common. The nine hallmarks and the seven SENS categories overlap substantially—to take one example, what de Grey calls "DNA damage" corresponds to "genomic instability" (the genome being the name for all of our DNA) in the hallmarks, which is a similar (albeit broader) concept. They also agree that there isn't a one-to-one correspondence between causes of aging and diseases. Most of the outcomes of aging, whether cancer, dementia or graying hair, can't be pinned on a single underlying biological factor, instead emerging as a consequence of several, both acting simultaneously and interacting with one another. Thus, as we examine them throughout this chapter and the rest of the book, I'll try to link diseases and symptoms to individual causes of aging, but things won't always fit neatly into one category.

Diseases aren't always the responsibility of a single actor, and sometimes we don't know exactly what underlies a given ageassociated problem at all. It might turn out that some hard-to-attribute problems are fixed by accident when we alleviate a given process of aging—or it might be that they illuminate new fundamental causes of aging we've missed. As science uncovers more about the aging process, we may well add more phenomena to this list—but, for now, there's plenty to get our teeth into.

As well as hopefully having direct medical application, the process of intervening in these hallmarks is one of the best ways to further our understanding of aging: if we eradicate one of them and find it doesn't make much difference to lifespan, then perhaps it isn't a root cause, or maybe something else kills us before it gets the chance; if we fix something and another item is fixed along with it, then it will elucidate the connections between these phenomena.

The first step is to talk about what these underlying

features of aging are. I've grouped mine into ten categories which I, too, shall call "hallmarks" of the aging process. They're very similar to the 2013 hallmarks (I've added two to the list, and grouped two together, meaning I've got one more hallmark overall), and follow the same rules: they increase with age, aggravating them worsens health and ameliorating them improves it.

Let's take a tour of them, starting with life's most fundamental molecule.

1. TROUBLE IN THE DOUBLE HELIX: DNA DAMAGE AND MUTATIONS

Inside most cells of your body lie two meters of DNA, an instruction manual of six billion molecular letters—A, T, C and G, known as bases—which contains all the information needed to build you. Incredibly, in spite of its two-meter length, it's squeezed into a nucleus just a few millionths of a meter across. DNA's double-helix shape is the most famous molecular structure in the world. It adorns everything from biology textbooks to company logos, serving as a visual shorthand for "science." But the platonic ideal of DNA, a pair of elegant intertwined spirals, a pure, pristine carrier of genetic information, belies the chaos in which it finds itself inside our bodies.

Crammed tightly into the nucleus, jostling with all kinds of other molecules, DNA is under a constant chemical assault which could damage its structure, or introduce typos into our genetic instructions. There are lots of ways DNA can be damaged. Perhaps the most obvious is external influences: toxins and carcinogens (the name given to anything which causes cancer) from food, cigarette smoke or nasty chemicals can inveigle their way into the nucleus and wreak havoc; UV in sunlight and radiation like X-rays or natural radioactivity can alter DNA, or even snap it in two. However, most of the damage is self-inflicted: chemical side effects of normal metabolism, the collection of processes through which food is turned into energy. It's estimated that every cell in your body suffers up to 100,000 assaults on its genetic code every day.

On top of that, every time a cell divides this entire genetic code has to be duplicated. Thanks to the incomprehensible number of cells in your body and their fast rate of turnover, over your lifetime you'll produce a couple of light-years of DNA—enough to stretch halfway to the nearest star—in the form of ten quadrillion near-perfect copies of your two-meter personal genome. Even the highest-fidelity copying and proofreading systems nature can devise will make occasional mistakes given that job spec.

Most forms of DNA damage are reversible because the cell can tell that something looks wrong and mend it—for example, there might be a molecule stuck to the DNA when it shouldn't be, and your cell's molecular machinery can chop it off. Perhaps the more troubling thing that can happen to DNA is that the repair process can go wrong, causing a mutation. Mutations change the information the DNA carries, altering the code of A's, T's, C's and G's which make it up, in a way which is indistinguishable from any other piece of DNA. Given the four molecular letters which make up DNA, a short piece of code could read GACGT. After a mutation, this might read GA**T**GT instead, but there's no way for the cell to "know" that there's anything wrong. That means that they can persist indefinitely, even if the code change is potentially harmful to the cell.

The most infamous consequence of accruing mutations is, of course, cancer. All it takes is a single cell to happen upon the wrong combination of alterations to its DNA, and the result can be an unlimited ability to proliferate which can allow it to grow into a tumor and ultimately be deadly. However, we also think that changes to cells' DNA which don't take them all the way to becoming a tumor can cause problems: typos in our cellular instruction manuals mean that cells don't behave as they are supposed to. This means that mutated cells can become dysfunctional over time, or it can cause them to become **more** functional in ways that act to the detriment of the body as a whole, in so-called "clonal expansions" which we'll talk more about in Chapter 7.

One piece of evidence for the importance of DNA damage and mutations is that people who are successfully treated for cancer in youth often end up with what is basically accelerated aging. The tragic and underappreciated shadow cast by our incredible successes bringing childhood cancers under control is an adulthood with increased risk of heart disease, high blood pressure, stroke, dementia, arthritis and even a higher likelihood of subsequent cancer, the end result of which is a reduction in life expectancy of about a decade. This is thought to be because many treatments for cancer work by damaging DNA. While chemotherapeutic drugs are carefully designed and radiotherapy's X-ray beams carefully pointed to ensure that the tumor bears the brunt of the damage, other tissues are inevitably hit by these treatments, too. The effect is highly specific: women who receive radiotherapy for cancer of the left breast rather than the right tend to suffer from more severe heart disease, because the accidental dose of radiation to the heart is unavoidably higher in these cases. This suggests that DNA damage and mutations can directly accelerate the aging of the heart—and implicates these processes in the broader phenomenon of aging too.

2. TRIMMED TELOMERES

If you know one thing about aging biology it's probably that it has something to do with telomeres. It does—but their story is a bit more complicated than how it's often portrayed.

The tale of telomeres starts out deceptively simple. Our DNA is split into 46 lengths called chromosomes (we get 23 from each parent). Telomeres are protective caps on our chromosomes, and their purpose is to solve two rather ridiculous evolutionary problems. First, they stop the flailing ends of our chromosomes from being mistaken by over-enthusiastic DNA repair machinery for loose pieces of broken DNA, and gluing them "back together," creating unintentional chromosome spaghetti.

Second, even more absurdly, our DNA replication machinery can't copy all the way to the end of a DNA molecule. You can imagine it like a builder who shuffles along the top of a long wall, building brick by brick as she goes—but, because she's got to stand somewhere, she can't lay the final bricks at the very end because she'd need to lay them under her own feet. This means that a small amount of DNA is lost from the end of a chromosome every time a cell divides. It would be untenable to lose important genetic information every time a cell divided—genes near the end of chromosomes would simply disappear, lopped unceremoniously off during DNA replication. Telomeres are evolution's answer—just make the code at the end of chromosomes be something trivial whose loss will be no tragedy as far as the cell is concerned. Thus, our telomeres are made up of hundreds or thousands of repeats of a six-letter sequence, TTAGGG, TTAGGG, TTAGGG, as far as the eye can see. When some of the telomere is lost when DNA is duplicated during cell division, nothing terrible happens.

It won't have escaped your notice that telomeres are a somewhat temporary reprieve from this problem. Losing a bit of DNA each time a cell divides isn't an issue when you've got long, youthful telomeres to burn through, but, as cells divide repeatedly and telomere length dwindles, so we get perilously close to truncating DNA that's actually important. As a result, when telomeres become critically short, they send out alert signals which stop a cell from dividing. After too many divisions, a cell will either commit suicide via a process known as apoptosis* or remain alive but stop dividing, in a state called senescence (which we will return to shortly—senescent cells are another hallmark of aging).

Every time a cell divides, around a hundred bases of DNA are lost. Cell division is an essential part of life for many of our tissues—for example, our skin constantly loses dead cells from its outer layers, and new skin cells divide afresh to replace them from beneath every few weeks—so our telomeres tend to shorten as we go through life. Telomere length is often measured in white blood cells, just because taking a blood sample is a simple procedure. A fresh white blood cell in a newborn baby might have telomeres which are 10,000 bases long (so about

---

* Pronounced a-puh-toe-sis, with a silent second **p**, thanks to being derived from Greek—like "pterodactyl."

1,700 TTAGGGs); by the time you're in your thirties, this will drop to 7,500 bases; by your seventies, the average telomere could be below 5,000 bases in length. This process is known as telomere attrition.

Short telomeres are found along with many of the diseases and dysfunctions of aging—they've been linked to diabetes, heart disease, some kinds of cancer, reduced immune function and lung problems. Telomeres are also implicated in the rather more superficial phenomenon of our hair turning gray as we get older. Stem cells in our hair follicles are responsible for producing melanocytes, the cells which produce the pigment melanin whose presence in varying quantities can make your hair anything from blond to black. When the stem cells' telomeres get too short, that means no more melanocytes can be produced, and the hair reverts to its "natural" color—pure white.

Short telomeres are also bad news for risk of death overall. A study looking at same-sex twins found that the twin with shorter telomeres was more likely to die first. The largest collection of telomere length data so far performed—looking at 64,637 Danish people—found that those with the longest telomeres were at 40 percent less risk of death than those with the shortest, even after their age and other factors affecting their health were taken into account.

Finally, our cells keep an eye on their telomeres for reasons other than their dwindling length. Telomeres are unusually susceptible to DNA damage, and there's emerging evidence that they act as a kind of canary in the coal mine for the rest of the genome—if a cell's telomeres have taken heavy damage, it's an indication that the rest of your DNA might be in a sorry state, too. Like telomeres that are critically short, damaged telomeres can signal to a cell that it's time for apoptosis or senescence. This is especially relevant in places like the heart and brain where we think cells don't replicate very often (or maybe at all) during our lifetime, meaning that their telomeres won't get shorter thanks to cell division—but where damage to telomeres, gradually accrued throughout life, can have similar effects.

Thus, through their length and their condition, telomeres are indicators of the health and history of a cell, providing a running report on whether or not a cell is aging well—and are therefore critical players in how we age.

3. PROTEIN PROBLEMS: AUTOPHAGY, AMYLOIDS AND ADDUCTS

We are protein. Though DNA often seems to get all the press, it's only the instruction manual. The instructions in DNA specify how to build proteins—molecules which are far more varied, far more complex, and do far more of the work.

The most immediate association of the word "protein" is probably the nutritional information on the side of packets of food, but to imagine protein as an amorphous nutrient like a bag of sugar or block of fat does this cornucopia of chemicals an enormous disservice. Proteins are the most diverse, intricate and complex molecules we know of: they're nature's nanobots, tiny, tireless molecular machines which keep us alive, and they're our cells' and bodies' scaffolding, the structural and mechanical building blocks which hold us together and allow us to move.

Autophagy

Many proteins have a short lifespan. An individual protein molecule, hard at work inside a cell, will typically last a few days. This might sound wasteful, the ultimate in throwaway living, but it's actually a huge advantage when it comes to aging, and bodily integrity in general. It's **because** they're so important that proteins are disposable: rather than investing valuable resources in making them indestructible, or devising ludicrously complex ways to fix thousands of molecules each of which can go wrong in uncountable ways, evolution has decided that it's usually best just to bin the broken protein and make a new one. Our cells are masters of recycling, chopping up old or damaged proteins into pieces which can then be reused in the next round of protein production.

One of the key processes involved in recycling proteins is "autophagy." Literally translating as "self-eating," autophagy is a way that cells get rid of rubbish—mangled molecules and broken-down cellular components which are no longer working correctly—and recycle their ingredients to make fresh new versions. Its importance to our cells' functioning was underlined by a Nobel Prize in 2016, awarded to Japanese scientist Yoshinori Ohsumi for his discoveries about how autophagy works.

Damaged cellular components, including many broken proteins, accumulate as we age, which is probably both a cause and an effect of a decline in autophagy with age. Reducing or entirely disabling autophagy in the lab can accelerate aging in worms, flies and mice. We also think that it's one of the mechanisms behind dietary restriction: disabling autophagy stops DR from extending lifespan, which suggests that it plays a key role. When food is scarce, autophagy frees up materials locked in existing proteins—with the added bonus that it tends to go after the broken stuff first, depleting damaged proteins and thus slowing aging.

We also know that age-related diseases can be triggered by problems with autophagy. One example is Parkinson's disease, a degenerative brain condition which causes sufferers to lose control of their movements: symptoms include rigidity, tremors and difficulty walking—and, in extreme cases, a total inability to move and wider symptoms of dementia such as difficulty thinking and emotional problems. Parkinson's patients have a life expectancy of about a decade after diagnosis, eventually dying from a variety of problems caused by loss of control of their muscles.

The risk of Parkinson's is increased if you have a mutation in a gene called **GBA,** which codes for one of the digestive enzymes involved in autophagy. Parkinson's is accompanied by "Lewy bodies," clumps of a protein called alpha-synuclein that are toxic to brain cells. The problematic, sticky form of alpha-synuclein is normally degraded by autophagy, but even a small impairment caused by a minor **GBA** mutation is enough to slow its breakdown, increase its levels and thus increase the risk of getting Parkinson's. Impaired autophagy is also associated with Alzheimer's and Huntington's disease, arthritis and heart problems.

Thus, the failure of autophagy with age, its association with age-related diseases and the fact that reducing or disabling it can cause diseases and stop life-extending interventions from working suggests that autophagy (and protein recycling in general) is an important part of the aging process.

Amyloids

In proteins, function follows form, and every protein's unique, intricate structure allows it to turn its hand monomaniacally to one highly specific task. The way that proteins acquire their incredibly complex, precise shapes is by folding—a kind of molecular origami which starts out with a long chain and bends and shapes it into everything from sheets and spirals to precise molecular keys which will only fit in the very specific lock provided by another protein.

Unfortunately, the exquisite complexity of protein folding means that even the tiniest fumble in this process can cause a protein to fold in a totally different way. One particularly nasty type of misfolded protein is the amyloid. These misshapen molecules can clump together, glued by sticky sections exposed by their misfolding. If you get enough amyloids together in the same place, they can form structures known as "amyloid plaques," which can strangle cells and tissues.

The most famous amyloids and amyloid plaques are those associated with Alzheimer's disease. The "amyloid hypothesis" suggests that a particular type of misfolded protein called amyloid beta is the prime mover in the disease, and that the molecular and cellular carnage which characterizes the later stages is all set in motion by the conspicuous aggregates of amyloid beta which develop in the space between cells in the brain. After decades of study and the failure of a number of amyloid-clearing drugs to help Alzheimer's patients, this ordering of events is now controversial and the amyloid hypothesis is coming under increasing pressure.

However, Alzheimer's is far from the only disease where amyloids are found—the alpha-synuclein aggregates of Parkinson's we just met are amyloids, too, and there are now dozens of diseases where amyloids are known to be implicated, from other brain diseases, to heart problems, to diabetes. These disease-associated amyloids are something which aren't found in either young or older healthy brains and blood vessels, so we're likely to need some anti-amyloid weapons in our anti-aging arsenal.

Adducts

Misfolding and aggregation into amyloid is one way a protein can go wrong. Another is for a protein to be made correctly, fold up fine, but then to have its structure modified afterward in a problematic way. In many cases, such a modified protein would be broken down and recycled via autophagy. However, some proteins are not rapidly renewed and replaced—they can live for months, years, or sometimes even as long as we do, meaning that the proteins themselves can age.

One of the challenges of being alive is simple chemistry. Fueling the many processes which keep our bodies ticking over requires us to have chemicals like sugars from our food, and the oxygen that reacts with it to release energy, floating about the place. No matter how cleanly you live, these highly reactive molecules are unavoidable, and they're a danger to everything around them—not least to proteins. Sugars are very keen to glue themselves to proteins in a process called glycation, and oxygen can do the same in reactions known, logically enough, as oxidation—and these additions to proteins are collectively referred to "adducts."

You probably encounter glycation every day—it's one of the most important reactions in cooking, thanks to a family of protein–sugar interactions known as the Maillard reaction. The Maillard reaction is behind the crust on bread as it bakes in the oven, the seared surface of a pan-fried steak and the aromas, flavors and dark brown color of roasted coffee. Unfortunately, the reactions which give rise to many of the most delicious flavors in food and drink are bad news for your body.

After a host of complex intermediate reactions, the final stage in the chemical bonding between proteins and sugars is an advanced glycation end product, or AGE. AGEs, along with proteins damaged by oxidation, are more or less irreversibly broken. Since the structure of proteins is so intimately related to their function, altering it by sticking sugars and oxygen on the side can impede them in their work, or change the way they interact with the proteins and cells around them.

This is mainly a problem for proteins which find themselves outside your cells, and glycation, AGE-ing or oxidation can affect different proteins in subtly different ways. Collagen, a structural protein with roles as diverse as skin flexibility and bone strength, can lose that strength and flexibility; the crystallin proteins which make up the lenses of your eyes can also become stiffer, making it harder to focus on nearby objects and meaning that almost everyone eventually needs glasses to read, and later for everything. Modifications to crystallins can also impact their transparency, leaving them clouded and ultimately causing age-related cataracts. Probably the most severe consequences come from stiffened blood vessel walls, thanks in part to modified collagen and another protein called elastin—this results in high blood pressure, which increases the risk of heart failure, kidney disease and even dementia.

Since many of the modifications we've discussed are made of sugars, their formation is accelerated if there is more sugar around. This means that diabetes can increase their numbers and worsen their effects. We often think of diabetes as a disease of high blood sugar, but it's the downstream consequences of sugary blood which are responsible for the worst side effects of the disease. Diabetics have significantly increased risk of heart attack and stroke, massively increased risk of kidney failure, and suffer nerve damage which can result in loss of sensation in feet and legs—at its worst, it can even make patients unable to notice when they're having a heart attack. Some of these symptoms are caused by glycation of proteins, bathed in a far higher concentration of sugar than normal at all times, and some is caused by the reaction of cells which aren't evolved to function in such a sugary environment.

Together, slowed recycling, sticking together as

amyloids and an accumulation of sugary and other modifications lead to problems with proteins that are responsible for many of the issues we experience as we age.

4. EPIGENETIC ALTERATIONS

Epigenetics is the collective term for a biochemical zoo of molecular decorations sprinkled on the DNA inside cells. It is a chemical code of its own which sits above (hence "epi") our genetics. Epigenetics unravels a seeming paradox in our biology: the cells of our body are almost ridiculously diverse, and yet almost all of them contain exactly the same DNA. Not only are there hundreds of different types of cells—skin cells, muscle cells, brain cells and many others—but those cells need to do different things at different times to ensure that they are responding appropriately to cues from your body, the environment and so on.

If your DNA is an instruction manual to build you, it's a particularly well-thumbed one, covered in bookmarks, placeholders and notes scrawled in the margin. These epigenetic annotations tell the cell what to do with the DNA they're attached to—whether, for example, to read a particular gene to be used in that cell at that time, or whether to ignore a whole section because it's never going to be needed.

There are dozens of different types of epi-genetic marks, but we'll concentrate on one of the best-studied in the context of aging: DNA methylation, meaning "methyl groups" made of a carbon and three hydrogen atoms that stick to your DNA. It's been known since the 1980s that DNA methylation tends to decrease overall with aging, but it was only with the sequencing of the human genome in the late nineties and the development of special "chips" which could measure methylation at tens or hundreds of thousands of locations across the genome that methylation could be understood in more detail. It turned out that our epigenetics knows how old we are even better than we do.

Steve Horvath, a mathematician-turned-biologist at the University of California, Los Angeles, was fascinated to know if patterns of DNA methylation could be used to glean any insight into aging. Unfortunately, very few people were specifically interested in epigenetics and aging at the time, but Horvath had an ace up his sleeve: the long-standing culture in genomics of making data freely accessible. Thanks to methylation chips being cheap and readily available, there were thousands of epigenetic datasets available from studies looking at other things entirely. Horvath combed through these, grabbing those which fulfilled one simple criterion: that the experimenters had made a note of the age of the patient the methylation had been measured in.

This sounds ludicrous, even with hindsight. The 8,000 samples he used in his first paper came from wildly different studies, looking at everything from diet to autism, preeclampsia to cancer, from different labs with different protocols and practices, and from different places in the body: blood, kidney, muscle, more than thirty different tissues and cell types in all. How could you hope to find anything in this haystack of disparate data?

He sifted through tens of thousands of methylation sites, and found just 353 which, together, were sufficient to predict someone's age. With this comparative handful of locations, the predictions were unnervingly accurate. The correlation between the predicted "epigenetic age" and the actual age was 0.96—where 0 would mean that they were totally unrelated, and 1 is perfect. This is off-the-charts performance: using telomere length to predict age, for example, scores less than 0.5. If you were to have your epigenetic age measured by Horvath's methylation clock, it would probably differ from your chronological age by less than four years.

This level of performance was so outlandish that Horvath's paper was rejected. The peer reviewers simply refused to believe that this ridiculous clock—built on a hodgepodge of data cobbled together from online databases, narrowed down to a tiny number of methylation sites—could possibly make such precise predictions **in any tissue of the body.** Horvath did eventually manage to get the paper published—though he later told a reporter that he had trouble believing the results himself until they were independently verified by other researchers.

The next step was to study people whose epigenetic age differs from their chronological age. Say you were actually 50 but your epigenetic age was 53, you'd be said to have an epigenetic "age acceleration" of three years. Multiple studies have now shown that epigenetic age acceleration is bad news—people with an epigenetic age beyond their years die sooner. Happily, the converse is also true—it's possible to be biologically younger than your calendar age, too, and thereby healthier and at less risk of death.

The morbid precision of epigenetic clocks suggests that either epigenetic changes are a cause of aging, or at least that they are a window through which to understand how our bodies get biologically older with time.

5. ACCUMULATION OF SENESCENT CELLS

As you look in the mirror each morning, aside from the odd fresh pimple, your face probably looks pretty much the same as it did the previous day. But your mirror is lying to you: our relatively constant external appearance from day to day belies a microscopic tumult under our skin, and throughout our bodies. The numbers sound terrifying—hundreds of billions of your cells die every day. Thankfully, you barely notice: first, you have a sum total of around 40 **trillion** cells, meaning the casualties number a tiny fraction of your total cell count; and second, the dying cells are constantly being replaced. This whole process is cell turnover, and its seamless operation is an essential part of life as a long-lived multicellular organism.

The cleanest end to a cell's life is the process of apoptosis, or programmed cell death, that we met above. A suite of molecular checks and balances is always keeping an eye on how an individual cell is behaving and, if something looks wrong, will initiate a tightly choreographed self-destruct cascade. The vast majority of frail cells do indeed do the decent thing for our bodies and die on cue, but some cells persist. They stick around, no longer dividing—aged, zombie cells which refuse to commit cell suicide, known as "senescent" cells.

This state was discovered in 1961, by a young scientist named Leonard Hayflick. He noticed something strange when growing cells in a dish: older cells looked noticeably different from younger ones and seemed, after a certain point, to stop dividing. This phenomenon was christened "replicative senescence"—the cells stop dividing because they've already divided too many times. The quantification of this phenomenon now bears Hayflick's name: the number of times a cell can divide before becoming senescent is called the Hayflick limit.

Hayflick's experiments overturned half a century of dogma which held that cells were immortal outside the messy confines of the body. Thus, an obvious question arose: does the senescence of cells contribute to the senescence of the organisms which they make up? Do we get old because, after a certain number of divisions, our cells lose their capacity to proliferate?

Three decades after Hayflick's work, we discovered that a hallmark we've already met—critically short telomeres—is the underlying cause of replicative senescence. We've also found that there are more reasons a cell can become senescent. One key driver is DNA damage and mutations—if there are enough pockmarks in a cell's DNA, especially in particular genes which put the cell at risk of turning cancerous, senescence will apply the brakes. Cells can also turn senescent when put under chemical or biological stress, which could serve a similar purpose—stress can induce cellular damage which, again, can be the first step on the road to cancer.

Thus, cellular senescence exists as an anti-cancer mechanism. Given that cancer is a disease caused by cells which divide out of control, turning a pre-cancerous cell into a senescent one which can no longer divide is a safe way to extinguish the spark before it flares into a tumor. Got pre-cancerous mutations? Under potentially carcinogenic levels of stress? Or just divided suspiciously many times? Better to go senescent to be on the safe side. However, it's not good enough to just sit there: as a senescent cell, you are no longer performing the function of a fresh, functioning cell in the tissue you find yourself in. So step two, having turned senescent, is to call for help.

Senescent cells do that by secreting inflammatory molecules which alert the immune system to their presence, asking to be removed. This means that an immune cell on search-and-destroy duty will be attracted over by the molecular ruckus and engulf the senescent cell, ridding your body of the problem. This molecular flag-waving is the SASP—the "senescence-associated secretory phenotype" ("secretory" because the cells secrete these molecules, and "phenotype" is a biological term meaning an attribute or behavior).

Perversely, it's this calling for help which provides the means for these cells to cause such damage in our bodies. If the SASP does catch the attention of a passing immune cell and secures the senescent cell a hasty removal, all is well—but, if the senescent cell survives and keeps on pumping out inflammatory chemicals, this can effectively accelerate aging throughout the body. Estimates of how many senescent cells there are in old animals suggest they are pretty few in number—only a few percent of cells turn senescent, even in very old animals or people. This doesn't seem enough to cause problems by directly compromising the function of tissues—but the SASP's inflammatory molecules can allow a handful of cells to act as the proverbial few bad apples. Accordingly, it seems to take very few of these cells to cause problems: one study found that injecting just 500,000 senescent cells into young mice—around 0.01 percent of their total cell count—was enough to cause physical impairment.

When we're young, the smallish number of sporadic senescent cells generated all around the body are largely dealt with by the immune system. However, as you get older, various processes cause these cells to snowball in number. First, the formation of senescent cells increases: as you age, your cells have divided more often, had time to accrue more DNA damage and exist in the more stressed-out environment of an aging body. Simultaneously, a weakening immune system is less able to find and eradicate the ballooning senescent cell population. In a final ironic twist, the SASP from existing senescent cells can actually breed more of them, in a deadly vicious circle.

The result of this snowballing is a heightened risk of many different diseases. Senescent cells are a smoking gun, often found loitering suspiciously near age-related diseases. A retinue of senescent cells accompanies tumors in cancer; heart disease, kidney disease and liver problems; the brains of people with neurodegenerative conditions like Alzheimer's and Parkinson's; the painful, swollen joints of osteoarthritis; cataracts, which cloud the lenses of our eyes as we get older; and the age-related decline in muscle mass, sarcopenia.

The list of diseases where cellular senescence appears to play a part is long, and growing as biogerontologists take an increasing interest in these zombie cells. It seems there is a strong evolutionary trade-off at play: so deadly is the specter of cancer to multicellular organisms that evolution is prepared to put us at risk of disease and deterioration in old age, just to be sure we won't get cancer in youth. This is a classic case of antagonistic pleiotropy, and a strong contender for a cause of aging: few and probably on balance helpful in the young, more and bad news in the old, and plausibly responsible for multiple diseases.

6. POWER STRUGGLE:
MALFUNCTIONING MITOCHONDRIA

Inside our cells roam herds of thousands of tiny beasts called mitochondria. They are how cells make their energy, and thus they're often described as "the powerhouse of the cell"—so often, in fact, that it's not just a cliché, but it's even slightly hackneyed to point out that it is one. Given the centrality of energy to the process of living, it's probably little surprise that mitochondria are implicated in the process of aging, too.

Mitochondria are extremely odd. They're often portrayed as a population of largely independent, bean-shaped objects—but things are more complicated than that. We now know that they frequently undergo "fusion" and "fission" events, joining together in anything from small teams to, occasionally, a single megamitochondrion which hangs like a cobweb around a cell's interior, and at other times break apart to go their separate ways. They're also the only part of your cell outside the nucleus to have their own DNA, stored in up to ten separate circular chromosomes per mitochondrion.

Mitochondria change quite substantially with age. There tend to be fewer of them in older animals' cells, and those mitochondria produce less energy. This reduction in mitochondrial number is related to risk of illness and death: people with the least mitochondrial DNA in their cells (which is used as a proxy for the number of mitochondria) are more likely to be frail and 50 percent more likely to die than those with the most. Just like the DNA in our cell nucleus, mutations in mitochondrial DNA increase with age in both animals and humans. There's also a mitochondria-specific type of autophagy known, logically enough, as "mitophagy," and guess what? It declines with age, meaning that messed-up mitochondria accumulate.

In terms of specific age-related diseases, mitochondrial fingerprints are found in parts of the body where energy expenditure is high. Muscles are one tissue which burns through loads of calories: damage to mitochondria is part of the process which leads to the loss of muscle mass and strength as we get older. They're also critical in the brain: in spite of making up just 2 percent of our bodyweight, our brains consume around 20 percent of our energy. That means its mitochondria are always running at full power and, accordingly, dysfunctional mito-chondria show up in diseases like Parkinson's and Alzheimer's.

In the lab, mice have been bred with particular mitochondrial defects which cause changes that seem, superficially at least, like accelerated aging. In "mitochondrial mutator" mice, a gene needed to copy mitochondrial DNA has been altered such that it no longer performs "proofreading"—meaning it doesn't check that the copy it's made is correct, and the mice accumulate lots of mutations in their mi-tochondrial DNA. They experience premature hair graying and hair loss, hearing loss, heart problems, and reduced lifespan. Another experiment bred mice with a mutation which reduced mitochondrial numbers, and could be turned on and off by giving the mice a drug and later withdrawing it. Activat-ing the mutation by giving the mice the drug caused them to develop thickened, wrinkled skin, hair loss and lethargy, rather like old mice; stopping the drug and giving the mice a few weeks to recover caused their wrinkles to fade and restored their fur to a condition indistinguishable from their litter-mates without the mutation.

The first theory to specifically implicate mito-chondria in the aging process was the mitochondrial free-radical theory of aging. Because mitochondria generate energy, they are forever dealing with highly reactive chemicals, particularly oxygen. If a mito-chondrion slightly fumbles the incredibly complex chemistry behind producing power safely, it can create a "free radical"—a highly reactive chemical which can go on to cause chaos in the cell, damaging proteins, DNA, and any other critical molecules it comes across. Three of the most biologically important free radicals are OH, NO and ONOO⁻: with names like that, surely these are prime suspects in causing cellular damage.

We now know that the simple idea of mitochondrial free radicals as biochemical berserkers which accelerate aging is an oversimplification. If aging was down to mismanaged free radicals, the killer experiment would be to increase animals' inbuilt defenses against them and watch as lifespans lengthened. Mice genetically engineered to have extra copies of anti-free-radical genes live no longer than normal mice. Worse, doing the opposite makes no difference either: an experiment in worms deleted all five of their anti-free-radical genes, called **sod** genes, and the worms suffered from massively increased free radical damage, but their lifespans were unaffected.

Recent studies have uncovered that free radicals are part of the extensive molecular vocabulary that cells use to communicate, and to regulate behavior. They tell cells when to grow, when to stop, and orchestrate processes like apoptosis and cellular senescence. Immune cells use them as weapons, overwhelming invading bacteria with an onslaught of free radical damage. Life has been dealing with free radicals for literally billions of years: with hindsight, it's obviously simplistic to assume that evolution would leave our cells at their mercy.

This new, more nuanced understanding of free radicals doesn't let them off the hook entirely: they still **can** damage our essential biological molecules, and it seems unlikely that this has literally no effect. Mitochondria are central to processes from cell growth to cell death and, as we've seen, mitochondrial behavior changes with age. Thus, mitochondria are key players in the aging process.

7. SIGNAL FAILURE

All around the body, our cells are constantly chatting, an endless exchange of molecular messages to the cells next door or on the other side of the body. This chemical telecommunications network has a huge impact on our physiology, from sex hormones, to sleep, to growth, to coordinating the immune system. All of these effects together are cell signaling—and, of course, as we age, they start to go off the rails.

Increasing dysfunction in signaling is part of the reason why aging all seems to happen at once, after decades spent in broadly good health. Because these signals wash throughout our bodies, dispersed in our blood, they can synchronize detrimental effects across our tissues. Even worse, it's a vicious circle—as our cells' condition worsens, aggravated by aged signaling, so the chemicals they secrete worsen things further. This downward spiral contributes to the exponential increase in the risk of death at older ages.

One of the major changes in signaling in our aging bodies is thanks to a process which will be a recurring character in this book, known as inflammation. Inflammation is our body's first line of defense against infection and injury, which often results in the site of the problem swelling up. The inflammatory response is the molecular equivalent of sending up a distress flare, calling out to cells in our immune system to rush in and fight the invaders, or to begin healing a wound. In youth, the normal process of inflammation is vitally important for ridding ourselves of infection and dealing with injuries. In old age, the inflammatory response can get itself stuck at too high a state of alert, which is "chronic inflammation" and can in fact fuel the aging process.

This process of gradually increasing inflammation as we get older is so ubiquitous that it's been dubbed "inflammaging." You can see it in blood test results: C-reactive protein, which doctors often measure to test whether you've got an infection, and interleukin-6, another molecule used to signal the immune system, are both known to increase with age. Not only that, but having high values of these inflammatory markers at a given age correlates with your risk of getting many of the diseases of aging which we're by now very familiar with—cancer, heart disease, dementia and so on. It seems that most age-related changes are aggravated by inflammation one way or another.

The reasons for this gradual increase in inflammation are manifold. We've already discussed one source of this spiraling dysfunctional signaling: senescent cells, and their toxic SASP. Some components of the SASP are precisely these immune-rallying molecules which contribute to an overall state of heightened alertness, and the level of alarm increases with the steady accumulation of senescent cells with age. There's also age-related damage, like the oxidized, glycated or otherwise broken proteins which we discussed: rather like with senescent cells, it's the immune system's job to clear up these damaged molecules. Eventually, as they become more numerous, there's a constant low-level hum of cries for help throughout the body. There are also persistent infections which our bodies can keep under control, but never quite eradicate (more on those shortly)—again, this results in a constant, low-level background of immune hyperactivity, with detrimental results.

When it comes to signaling, there's also an intimate relationship between our body's response to food, and detrimental chemical messages associated with old age. This is known as "deregulated nutrient sensing," because our bodies lose the ability to sense nutrients and respond to their presence appropriately, and the key experiments underpinning this are the DR studies that we met in the last chapter. Part of this is "insulin resistance," which is the precursor to diabetes—when your body doesn't respond properly to the hormone insulin, which tells cells to pull sugar out of the blood and store it for later use. We all now live in an environment with easy access to food or drink, including many options which are very sugary, meaning that many of us are at risk of developing diabetes,* especially as we get older. In people with insulin resistance or diabetes, insulin becomes "the hormone that cried 'Wolf!'" and, even if more insulin is produced, its call to arms goes unheeded—the fat and liver cells that normally sequester sugar ignore the insulin signal and leave the sugar in the blood, where it can do damage.

However, diabetes isn't just caused by an over-familiarity with sugar and insulin—otherwise, unless we all developed a sweet tooth with age, why would it increase as we get older? We now know that insulin resistance and diabetes are also driven by inflammation. One piece of evidence connecting these is that patients with severe infections often develop rapid-onset insulin resistance and accompanying sky-high blood sugar due to the massive inflammatory response which is marshaled to fight them. In aging, chronic inflammation causes a similar process in slow motion.

---

* Specifically, type 2 diabetes, the form usually associated with aging. Type 1 diabetes is an autoimmune disease, where the immune system attacks the cells that produce insulin in the first place.

As well as inflammation and nutrient sensing, there are lots of other signals that go up and down in our bodies with time. They include other hormones like oxytocin, "growth factors" that tell cells when to proliferate and build tissue and when to hold off, and messages in a bottle called exosomes in which cells send little packages to both their neighbors and cells across the body. The wide-ranging changes in all these types of signaling with age mean that these messenger molecules are central to the aging process.

8. GUT REACTION: CHANGES IN THE MICROBIOME

There are trillions of microorganisms—bacteria, fungi and viruses—living in and on you right now. These hitchhiking bugs are collectively known as our "microbiome," and they live on our skin, in our mouths and, particularly, in our guts. The microbiome is a hot research topic at the moment because these microbes have turned out to be far more than just passive fellow travelers—they help with breaking down our food, work to keep us free of infection and even, fascinatingly, talk to and help out our immune system. Estimates of their numbers vary (not least with how recently you last went to the toilet—since most of them live in your large intestine, bowel movement can subtract as much as a third from your microbe population), but it's thought that the number of microbial cells in your gut is approximately the same as the number of human cells in your body. Given their sheer quantity, it's perhaps unsurprising that they can have a significant effect on your health.

One common theme of microbiome research is that there seems to be strength in diversity. A rich, varied population of gut flora is good news: when you're young, a diverse family of microbes in your gut helps out with digesting certain components of food, overpowers invasive bacteria like those that cause food poisoning and chats amiably with your immune system. As you get older, or if you have a chronic condition like irritable bowel syndrome, diabetes, colon cancer or even dementia, your intestines can come to be dominated by fewer types of microbe, and often more aggressive ones. The direction of causality here isn't entirely clear—it could be that the loss of diversity is driven by poor health or poor diet, or that a change for the worse in gut microbes impacts negatively on the rest of the body. Because this is biology, it's quite likely to be a bit of both.

One mechanism by which a dysregulated microbiome is thought to affect the aging process is by contributing to chronic inflammation. As diversity wanes and more aggressive microbes begin to take over, the immune system is put on high alert to keep these potentially infectious organisms in check. It's also thought that the lining of our guts gets a bit leaky with age, driven by both other hallmarks of aging taking their toll on the cells lining them and the changing microbiome. These leaky intestines allow a few symbiotic microbes, microbial toxins or tiny pieces of food to cross into our bloodstream, which again causes a low-level thrum of immune activity, worsening inflammation.

As well as the aging process itself, getting old is associated with other factors which drive changes in our microbiome. Our gut flora is heavily influenced by our diet, since the microbes effectively share what we eat. Older people's diets often change significantly, sometimes for trivial-sounding reasons, like consuming less fruit because it gets harder to eat as we lose our teeth. This is a beautifully simple example of why it's so important to treat the whole aging process, and not do medicine in silos—better dentistry doesn't just mean less toothache, but also has impacts on diet which have wider knock-on effects of their own. The elderly are also prescribed more antibiotics, which can wallop the microbiome at the same time as treating the bacteria making them ill. There's also a significant environmental component to your microbial family, and older people living in care often have a different spectrum of species from those still living at home.

In spite of this complexity, we have managed to build "microbial clocks" which, rather like the epigenetic clock we met recently, can determine someone's age to within four years or so based on the relative proportions of different bugs in their guts. There's also evidence from animal studies that the microbiome can be problematic in old age. One study took young and old mice without a microbiome of their own and placed them in cages with other mice, also either young or old. Mice living together exchange gut bacteria by "coprophagy"— a scientific euphemism for eating one another's poo—and the microbiome-free mice thus adopted their cage mates' bugs. Those which ingested old-mouse microbes saw an increase in gut leakiness leading to raised levels of inflammation throughout their bodies, supporting the hypothesis that bad old bugs actively worsen health in old age.

As a hallmark of aging, the microbiome is the most tentative on this list. There isn't currently a huge literature on the microbiome's effects with age, but studying our bodies' microbial ecology is an emerging discipline so we shouldn't expect to have all the answers yet. It's also phenomenally complicated, involving a fast-paced ecosystem of thousands of species of bacteria, fungi and viruses interacting with one another, our diet and environment, intestines and immune system, meaning that working out the details could take a while. However, the meteoric rise of microbiome research over the last decade or so means we can expect to know a lot more about it, and its effects on aging, in the next few years.

9. CELLULAR EXHAUSTION

It's probably no surprise that, besieged by the hallmarks of aging we've met so far, we start to lose cells in our body as we age, and those that survive become worn out and less able to do their jobs properly. The catch-all name for these processes is cellular exhaustion, and it affects many of the populations of cells in tissues and organs across the body.

Stem cells are the type of cell whose exhaustion is most often discussed—because it's their job to replenish when cells wear out, so if they themselves become exhausted, that's bad news. Stem cells have particular importance in places where cell turnover is high. For example, hematopoietic stem cells (a bit of a mouthful, so let's call them HSCs) reside in our bone marrow and work tirelessly to renew the various cells which make up our blood. Together, they are responsible for producing 200 billion oxygen-carrying red blood cells, plus billions more immune cells and platelets (which are used in blood clotting), every single day.

As we age, HSCs become less effective at replenishing our blood cells. This is down to a number of the hallmarks we've already discussed, including DNA damage and mutations, epigenetic changes, problems with autophagy and changes to signaling from cells in their environment. The irony is that all of these changes actually **increase** the number of HSCs overall—in part because these factors bias them slightly toward dividing to make two stem cells, thus increasing their own population, rather than into a stem cell and a blood cell precursor.

As well as producing too many stem cells and not enough of the type of cells they're meant to top up, stem cells can produce the wrong ratio of cell types with age, too. One example of this is mesenchymal stem cells (another mouthful—let's call them MSCs). These are a group of stem cells whose progeny include bone-building osteoblast cells, tissue-joining cartilage cells, muscle cells and a type of fat cell found in bone marrow. As we age, MSCs lose their taste for forming bone-builders, while increasing their preference for becoming fat cells. This means that our bone marrow has less of the strength-giving matrix of proteins and minerals which osteoblasts deposit, and more fat. Fattier bones are weaker, and this process contributes to osteoporosis—the age-related weakening of bones which is particularly acute in women after menopause. This weakens our skeletons, which is a problem often without obvious signs until you end up in the hospital with a serious fracture. It can also result in many tiny fractures which go unnoticed—repeated "compression fractures" crushing the bones of the spine are one of the reasons we get shorter with age. That means that problems ranging in severity from millions of broken hips to countless lost centimeters of height can be partly blamed on a change of preference among a particular kind of stem cells.

The decreasing effectiveness of stem cells has wide-ranging consequences throughout the body. Another thing we can chalk up to it is the reduction in our senses of smell and taste as we get older. Odors are picked up by a specialized group of brain cells called olfactory receptor neurons which protrude into a part of the roof of the nasal cavity. Tiny hair-like structures covered in receptors sample the molecules drifting into our noses, and send signals to our brain with news of what they've detected. Since they have to be in contact with the outside world in order to function, they inhabit an unusually traumatic environment for a neuron, under siege by environmental toxins and microbes rather than safely shielded inside the skull. Consequently, they die relatively frequently and rely on stem cells to replace them. Olfactory neuron stem cells start to flag as we get older—more of them sit idle and, unlike HSCs, they seem to develop a preference for dividing into non-stem-cell daughters, meaning that the pool of replenishers shrinks. Thus, as smells fade and food no longer tastes quite like it used to, the wearing out of stem cells is responsible.

Though stem cells are the group whose exhaustion is most often considered, aging is also evident in tissues which don't rapidly replace their cells. Tissues like the blood, skin and intestines are known as "renewal tissues" because they're constantly renewing. Where an intestinal stem cell might divide once a week, stem cells in the liver do so once a year. Some tissues, such as heart muscle and many parts of the brain, may not renew at all. This is why, when heart cells die after a heart attack or brain cells are lost after a stroke, the damage is often permanent.

It's via this subtly different mechanism that another sense is blighted by cell loss in old age: hearing. We hear when sound is channeled down our ear canals to our inner ears, where tiny hair cells detect the vibrations and send signals to our brains about what kind of sound it is. Unfortunately, loud sounds, toxins and simply growing older can all damage these hair cells, and we can't replace them. Older people particularly lose the ability to hear high notes, and general loss of sensitivity at all frequencies reduces the ability to clearly distinguish sounds and understand speech.

This is a blight on the lives of the elderly—the inability to hear is everything from socially isolating (imagine being unable to hear what everyone around the dinner table is saying) to outright dangerous (imagine crossing the road while being unable to hear the traffic). Cell loss without replacement is responsible. It also has an indirect effect, with the loss of sensory stimulation putting people with hearing loss at greater risk of dementia. Our current treatments are also textbook examples of attempting to plaster over the problem rather than treating the underlying cause: hearing aids simply cause sounds to be louder, allowing atrophied ears to make them out. Unfortunately, the indiscriminate amplification of a hearing aid doesn't have the finesse of our meticulously evolved aural system, meaning it can often be hard to do things like pick out an individual voice in a noisy environment when using them.

Thus, the loss either in numbers or in effectiveness of stem and other types of cells around our body is responsible for some of the slow decline of aging, and some specific diseases. This is likely to be driven by many of the hallmarks we've already discussed, but it's a major enough issue to warrant a hallmark of its own.

10. DEFECTIVE DEFENSES: MALFUNCTION OF THE IMMUNE SYSTEM

As we've just seen, the end result of everything from mutations to signaling problems is that our cells start to die or malfunction. The final stage in degeneration is the problems with organs and systems resulting from these failing component parts. From brain, to blood, to bones, to gut, every aspect of our physiology changes for the worse as we age. Many of these form vicious cycles, as we've already met with chronic inflammation: changing conditions in the body cause organs to work differently, often trying to compensate, which drives our bodies further from the relative stability they enjoyed in youth. One place where dysfunction and detrimental attempts at adaptation have particularly wide-ranging effects is the immune system.

The most obvious consequence of the immune system's reduced effectiveness in old age is that it is less able to defend us from infectious disease. This loss of defensive capabilities is clear from the statistics: considering the billion and a bit people who live in high-income countries, where vaccines and antibiotics are widely available, infectious disease is still responsible for a significant 6 percent of deaths. Our huge success with hygiene and modern medicine hasn't entirely eradicated the burden of infectious disease, but postponed it.

By alleviating mortality in childhood and young adults, many of us now live long enough to experience the immune decline of aging, and more than 90 percent of deaths from infectious disease are in people over the age of 60. The substantial extra risk to older people from infectious disease has been laid bare by the coronavirus pandemic, whose toll in terms of hospitalization and death is much higher among the elderly. While dying from flu or COVID-19 isn't aging per se, the massively increased risk with age means that aging bears ultimate responsibility for most of these deaths.

What's worse is that a key tool of modern medicine—vaccination—is less effective in the elderly because vaccines rely on the failing immune system for their effectiveness. By effectively giving immune cells a sneak preview of a potential disease, vaccines allow them to learn what to look out for. Unfortunately, as the immune system ages, our response to vaccines weakens, too. This doesn't mean it's not worth getting your annual flu shot if you're getting older—quite the opposite: because your risk of serious complications or death from flu is so much greater than when you were young, the overall protective effect of the vaccine is greater, in spite of a reduced immune response. (If you're young it's probably worth getting one anyway because flu is really quite a nasty disease—plus it will help to protect your older friends and relatives.)

Some of the reasons for this immune decline are other hallmarks of aging that we've already met. One key process is cell loss in a small organ called the thymus, which is found just behind your breastbone and in front of your heart. The thymus is the training ground for T cells (indeed, their thymic origins lend them their name), one of the two key types of immune cell which form the "adaptive" part of our immune system which is able to adapt to fight new threats.* The adaptive immune system can also learn—once it's fought off a particular threat, the victorious T cells can transform into "memory T cells," ready to ride again should the same germs

---

* The other main component of the adaptive immune system is B cells, which mature in the bone marrow. There's also the "innate" immune system, which comprises a range of generalist cells which can fight many different types of invader—including macrophages, which we'll learn more about shortly. There's not space to do justice to the kaleidoscopic diversity of the immune system, so T cells and macrophages will serve as our exemplars in this book.

return. Given how useful T cells are, it may come as a surprise to learn that your T cell academy is already mostly gone, unless you are reading this book at quite a precocious age. The thymus peaks in size at age one, and it's all downhill from there, halving in volume every 15 years or so—it's half gone by your teens, 75 percent gone by the age of 30, and barely any remains after the age of 60 or so. This disappearing act is "thymic involution," and it takes the form of previously functional thymic tissue turning into fat.

Though it might seem ridiculous, this process actually appears to be intentional. The evolutionary rationale for destroying your own defenses is that producing new T cells is expensive—and, as we know, it's often better to put energy toward reproduction rather than ensuring your own survival into old age. If you were a human living in prehistory, probably in a small group, unable to move further than your clan could roam, it's likely that you'll have seen most of the germs you'll ever need to battle by the time you're 20. This means you can save a lot of energy by producing fewer new recruits as time goes on, relying more on your memory T cells to keep you safe. It's a classic case of antagonistic pleiotropy and disposable soma theory, where freeing up resources to reproduce in early adulthood ends up costing us in late life—particularly today, as we now live decades longer in a highly connected world with constant exposure to new infections.

Memory T cells can stick around for decades and, if their old foe returns, they're the most highly proliferative cells in the body: a handful of memory cells can expand to form a clone army of millions. This puts the cells themselves under incredible strain—DNA damage and shortened telomeres from dividing many times can lead to cellular damage and senescence of the immune cells, which weakens our immune defenses.

The immune system also suffers its own peculiar forms of aging. The strangest is that the immune system can be aged by the very infections it fights: persistent bugs can lead to an immune obsessiveness which undermines its ability to face down new threats. Chief among these is cytomegalovirus, or CMV, a relative of genital herpes and chickenpox. The majority of people catch CMV at some point during their lifetime, and those infected never quite shake it off. As we age, T cells specialized in dealing with CMV can come to occupy up to a third of our "immune memory," leaving less "storage space" to learn how to deal with new infections.

And, though the immune system is most famous for seeing off external threats, it also has crucial roles to play in keeping internal ones under control. We've already seen how the immune system seeks and destroys senescent cells, and that its dysfunction in old age could be both a cause of their rise in numbers and exacerbated by their contribution to chronic inflammation. Immune cells are also on the lookout for cancer—trying to catch cells which are putting together the toolbox of genetic changes needed to form a tumor, but have nonetheless escaped senescence or apoptosis. Cancer's higher occurrence in old age is in part down to the immune system's decline: as its function declines with age, nascent tumors get more time to grow unchecked.

Another age-related problem which can trace its origins to the immune system is one you might not expect—heart disease. We hear so much about cholesterol and heart disease that you might imagine that our arteries are blocked with greasy deposits of cholesterol itself, but things are actually more complicated than that. The "plaques" responsible for heart attacks and much more besides aren't just streaks of lard, but graveyards of immune cells which died after becoming engorged with cholesterol. This process is atherosclerosis.

Cholesterol tends to get a pretty bad rap, because having too much of it in your blood can put you at risk of heart disease. In spite of this, it's actually an essential molecule in the body used, among other things, for building cell membranes—the bag which holds the contents of a cell together. The problem is that cholesterol often gets stuck in the arterial wall and starts a chain of events which can go on to kill you.

A plaque usually starts off with a small, innocuous injury. The alarm is raised, and immune cells in the blood rush to the rescue, engulfing whatever is causing the problem to make space for repairs to begin. Often, that's cholesterol. At first, the "macrophages" (literally "big eaters," a type of unfussy bad-stuff-gobbling immune cell) are quite effective at clearing up cholesterol. Unfortunately, though, it doesn't take much for them to become overwhelmed with more cholesterol than they can deal with. Worse, cholesterol can react with oxygen and sugars, just like the oxidized and AGE-d proteins we met, and macrophages can't deal with these modified versions. This means that they start to hoard rather than discard cholesterol, collecting it in fatty globs called lipid droplets.

This is stage one of atherosclerosis. The engorged, dysfunctional macrophages look foamy under the microscope, earning them the name "foam cells." Eventually, the foam party can overwhelm the macrophages, which results in them committing cell suicide. Guess what comes to clear up? More macrophages.

Of course, since the debris and dead cells include the same damaged cholesterol which finished off the previous macrophages, the new clean-up squad doesn't really stand a chance. As a result, they can die, too, setting up a vicious cycle. As more macrophages turn up and die, leaving an ever-larger pool of damaged cholesterol and dead cells, what was a microscopic injury becomes visible inside the artery as a "fatty streak."

Our first fatty streaks appear in our arteries as

children or teenagers, but it's usually decades before any of them reach the point where they pose a serious threat. Many years in the making, a full-fledged atherosclerotic plaque is an incredibly complex structure: at its core is a huge mass of dead macrophages and cholesterol, held in place by other types of cells which try to keep a literal lid on things.

This huge mass creates a bulge in the arterial wall which can narrow its interior, reducing blood flow. This is bad in itself, but usually doesn't cause an issue until almost the whole blood vessel is blocked. If that does happen, and it's a particularly critical vessel, it can cause serious problems—a blockage in one of the arteries supplying the heart, for example, can significantly reduce its supply of oxygen and lead to chest pains and shortness of breath. Nowhere in the body is safe from this narrowing: atherosclerotic plaques narrowing the arteries which supply the penis with blood can cause erectile dysfunction, by disrupting the surge of blood needed to get and maintain an erection.

The worst-case scenario is if the plaque ruptures: this causes it to spill its semi-solid contents into the bloodstream, where they can rapidly move through the body, jamming up smaller vessels entirely. If this happens in an artery supplying the heart, it will cause a heart attack—a total loss of oxygen to a part of the heart muscle. This will often be accompanied by chest pain, shortness of breath and a (justified) feeling of overwhelming dread. As is probably obvious, the victim should be rushed straight to the hospital; if the blockage is dislodged quickly enough, some of the affected section of heart muscle may be saved. The damage will almost certainly leave the heart weaker and, given the extremely slow turnover of heart muscle cells we mentioned, it is unlikely to ever fully heal.

Another particularly bad place for a chunk of plaque to end up is the brain. A blocked vessel here causes an "ischemic" stroke—a stroke resulting from a loss of oxygen to part of the brain tissue. If the blocked vessel is a big-ish one, this can cause immediate consequences such as weakness in face or arm muscles, loss of the ability to speak, blurred vision or dizziness; anyone with these signs should get to a hospital as quickly as possible. If it's a smaller vessel, then the consequences may be minor enough to go unnoticed but, over a few years or a decade, dozens of mini-strokes can accumulate to reduce brainpower and memory overall, in a condition known as vascular dementia. Stroke is a huge killer, responsible for around 10 percent of deaths globally. Even if it doesn't kill you, a stroke can leave you disabled, with difficulty moving, speaking and understanding, or partially blind. Vascular dementia, too, is a serious and widespread condition—the less infamous cousin of Alzheimer's is the second most common form of dementia, responsible for about 20 percent of cases.

Our immune system's inability to deal with

damaged cholesterol as it clears up debris in our blood vessel walls has a lot to answer for: depending on where you are in the world, the heart attacks and strokes caused by atherosclerotic plaques are behind something like one in five deaths. It's incredible that this single, obscure-seeming drama in our arteries is competing with cancer for the title of deadliest process in human biology.

FIXING THE HALLMARKS OF AGING

We've now met ten hallmarks of aging. Nothing in the body, from DNA to proteins to cells to whole systems of the body, is left unaffected by the ravages of time—which means we've taken a tour of much of biology. However, while you'd be forgiven for finding this list daunting, it's actually incredible that it's so short.

The human body contains something like one hundred organs, hundreds of different kinds of cells, and there are at least thousands of age-related diseases, depending on how you count all these things—and yet we can divide the villains behind aging into just ten categories. This means that we could potentially address many, perhaps most, of the changes and diseases we associate with aging by devising a far smaller number of treatments than our current approach to medicine—targeting every disease separately—requires. And we can start doing so right now.

Rather than going after hundreds of types of cancer and finding bespoke treatments for each, we could try to deal with the DNA damage which underlies them all, the senescent cells and chronic inflammation which aggravate them all and the faltering immune defenses which they must all slip past, and reduce the odds of getting cancer in the first place.

You may have noticed throughout the list of hallmarks that I only discussed two of the three criteria for a hallmark of aging: that it gets worse with age, and that accelerating it accelerates aging. That's because the next part of the book will look at how we can fix them. This isn't just a hypothetical exercise: what's most exciting is that ideas to tackle all of them are at various stages of development, from the lab to human clinical trials.

Part II will examine how humanity might turn the hallmarks into medicine, taking the first steps toward negligible senescence. We'll survey the treatments we currently envisage, based on undoing each of these hallmarks or making them irrelevant. It's split into four chapters which will look at removing bad things that accumulate, renewing things which are broken or lost, repairing things which are damaged or out of kilter and, finally, reprogramming our biology to slow or reverse aging.

Now that you know why we age, I want to show you how we can stop it.

PART II

Treating Aging

5

Out with the Old

Given that aging is a multifaceted process, treating it is going to require a portfolio approach. Unless we uncover some deeper root cause of aging, or it turns out that one or just a few of the hallmarks we examined in the last chapter are responsible for most of the problems associated with getting old, there may well be dozens of treatments in our eventual repertoire. The next few chapters will look at how we could intervene in each aspect of the aging process, with treatments that range from the currently hypothetical, to some nearly ready for deployment.

I've divided the treatments into four loose categories, each with a chapter of its own. The first is probably the most intuitive: the removal of bad stuff which accumulates with age. Some of the hallmarks of aging are simply things which build up in our body as we get older and go on to cause the diseases and dysfunctions of old age. Thus, we need to devise ways to get rid of them.

There are three hallmarks which follow this pattern: aged, senescent cells, whose numbers slowly increase with age; defective proteins and other junk that hangs around inside cells and slowly causes them to work less effectively; and misfolded proteins called amyloids which accumulate both within and between cells and gradually cause problems from heart failure to dementia. The most obvious way to tackle these problems is to remove them at their source.

We'll start with what's probably the strongest contender for first anti-aging treatment to make it to the clinic: removal of senescent cells.

KILLING SENESCENT CELLS

As we saw in the last chapter, senescent cells slowly accumulate in our tissues as we get older. Their name literally means aged cells, and senescence happens to cells whose telomeres are too short, that have suffered too much damage to their DNA, or that are just generally under catastrophic levels of cellular stress. As a result, they slam on the brakes and stop dividing for safety reasons. It's probably better than the alternative, which could be to turn cancerous, but the senescent state is far from benign: the cell pumps out molecules which fuel chronic inflammation around the body and can turn nearby cells senescent, too, or, ironically, cancerous. If not rapidly removed by the immune system, senescent cells fester, worsening both their local environment and the state of our bodies as a whole.

This shows us that senescent cells satisfy two of the criteria for a hallmark of aging: they accumulate with age, and their presence accelerates the aging process. There's only one further piece of evidence needed to clinch the case against them: does getting rid of them make things better?

The first evidence that it does was published by scientists working at the Mayo Clinic in the U.S. in 2011. It was a proof-of-concept study, which means it was unrealistic in a couple of key ways. First, it used mice bred with a genetic defect which makes them age more rapidly ("BubR1" mice), so the lessons learned would not necessarily map onto normal mice. Second, these mice were further genetically modified, and given an extra gene which would cause senescent cells to commit suicide when activated by a particular drug. Since neither normal mice nor human patients have this crucial engineered gene, they wouldn't have any reaction to the drug. You can drink as much of this elixir of life as you like, and it will have absolutely no effect.

But the results were clear. The first thing to check for was whether the mice had fewer senescent cells after the drug was administered, which they did. Far more excitingly, they also showed an improvement in a number of aspects of the premature aging normally experienced by this type of mouse. They had bigger muscles, and could run harder for longer on a treadmill; they had more fat under their skin (one of the reasons skin sags in older mice, as in older people, is loss of this so-called "subcutaneous" fat); they developed cataracts later than usual; and, careful biological measurements be damned— they just looked great, with plump, healthy-looking bodies and glossy, thick fur next to their hunchbacked, skin-and-bones counterparts who weren't given the drug.

The only thing which didn't improve was overall lifespan. That's because, even as they impinge on physique, appearance and quality of life, the disabilities caused by senescent cells aren't the ticking bomb which eventually ends a BubR1 mouse's life. Instead, this unfortunate breed of mouse dies of heart failure, which isn't much affected by the presence of senescent cells. Nonetheless, this study was a landmark—the first demonstration that removing senescent cells in an animal relieved the burden of age-related disease and dysfunction.

However, this result wasn't a slam dunk. Those caveats we mentioned—that this was done in genetically modified, fast-aging mice—means that this work only really caught the attention of a few senescent-cell geeks. It would be far more convincing in normal mice. A study published in 2015 by the same Mayo Clinic group rose to the challenge, cooking up a clever combination of drugs that goes after the senescent cells directly in regular, unmodified mice.

The scientists' strategy was to find a drug that nudged the cells toward suicide. It turns out that senescent cells harbour a deep ambivalence about their own continued existence: on the one hand, senescent cells are on their last legs due to damage or stress and really quite want to die, meaning that they activate many of the genes that promote pro-grammed cell death (apoptosis); on the other, they simultaneously activate genes which hold it back, keeping themselves alive. Maintaining the senes-cent state is a constant battle between the opposing forces of life and death. Perhaps it would be possi-ble to find a drug which would suppress the suicide suppressors and break the deadlock in death's favor?

The team identified 46 candidate drugs known to interfere with the anti-suicide genes and, after testing them all for senescent-cell-killing potential, identified two winners: a cancer drug used in che-motherapy called dasatinib; and a flavonol called quercetin which is found in fruit and vegetables, and sometimes taken as a dietary supplement. When used together they were even more effective, zapping senescent cells in multiple tissues in the body, while leaving innocent bystander cells unharmed. That means the Mayo team's "D+Q" cocktail was the first ever "senolytic"—a treatment which causes lysis (a biological term for falling to bits) of senescent cells.

The final step was to administer D+Q to aged mice. The results have been impressive: giving old mice D+Q basically makes them biologically younger. The 2015 study administered this cocktail to 24-month-old mice (equivalent to about 70 in human years), and found improved heart function and greater flexibility of blood vessels. An avalanche of subsequent work has looked for D+Q's effects in many other places: killing senescent cells in mice has been shown to improve atherosclerosis, Alzheimer's, diabetes, osteoporosis, regeneration in old hearts, lung disease and fatty liver disease; it helps old mice run farther and faster, strengthens their grip and lets them hang from a wire for longer. This list will undoubtedly be out of date by the time you read it—cellular senescence and senolytics are now incredibly hot topics, with new roles for these zombie cells being uncovered constantly, often by the simple process of removing them and watching how things improve.

Many studies concentrate on individual diseases because it's quicker to show a short-term change in a specific part of the body than it is to look at healthspan and lifespan overall. But a 2018 study showed that D+Q also has a global effect—a repeat prescription begun at 24 months helped mice live just over 6 months longer, versus 4.5 months without the drug. Even though treatment was started very late in life, these mice were living the equivalent of five or ten additional years in humans. Critically, this extra lifespan wasn't lived in a state of geriatric dysfunction; rather, aging seemed to be deferred. In post-mortems, the group of mice given the drug and those that died younger without it looked very similar—D+Q hadn't just delayed a single disease and thus extended lifespan, but slowed or partially reversed the aging process as a whole.

If started earlier in life, senolytics may have even more potential. A 2016 study used the old-school drug-activated genetic modification, but this time put the gene in otherwise normal mice, rather than BubR1 mutants. Starting the suicide-activating drug at 12 months, which is mouse middle age, delayed the onset of cancer and cataracts, improved the function of the heart and kidneys and even made the mice more curious in new environments than control mice not given the drug. It also increased average lifespan by 25 percent.

Given the range of contexts in which senolytic drugs have been shown to work, as well as the fact that they improve lifespan overall, senescent cells are clearly key players in the aging process—and getting rid of them will be key to treating it. The next step is to get senolytics working in people. Trials have already started.

The first human clinical trial was published in early 2019, a small safety study using D+Q in patients with idiopathic pulmonary fibrosis, or IPF, a disease in which lung tissue becomes badly scarred. IPF is thought to be driven in significant part by senescent cells, and D+Q in mice has been shown to improve it. This was an early pilot experiment, mainly concerned with checking that D+Q is safe to use in people, and involved just 14 participants. The results were positive—the drugs seemed safe, and there was even a modest improvement in physical performance, measured by how far and how fast patients could walk, and get up from a chair. Plenty more work remains to be done before IPF patients will be taking senolytic therapies, but this isn't a bad start.

While scientists from the Mayo Clinic continue to investigate D+Q, others are pressing ahead with their own senolytic formulations. The furthest along is probably a company called Unity Biotechnology, with two drugs known enigmatically as UBX0101 and UBX1967, targeting the knees of patients with osteoarthritis and the eyes of those suffering from age-related macular degeneration. These might sound like rather random places to start for drugs which delay a significant subset of the aging process, but there is a logic to them. First, it's better to begin with diseases where there's a specific suspicion of senescent cell involvement, where symptoms are clear and the effects could be seen relatively quickly. It's also better to start by treating people who already have unpleasant diseases; the inevitable side effects are more acceptable if your lungs are already in bad shape from a condition with no known treatment, as opposed to if you're an otherwise healthy middle-aged person. Finally, the knee and eye are advantageous because they're small, contained bags of fluid—you can inject a drug there and expect that only a minimal amount will leak into the rest of the body, reducing the likelihood of side effects. (Also, slightly more darkly, eyes are good because you have a backup—the other eye can act as both an experimental control and an insurance policy in case anything goes terribly wrong.)

The rate of progress in senolytics is dizzying. Going from the first proof-of-concept study in mice in 2011 to the first human trials starting in 2018 is incredibly rapid progress by medical standards. All being well, it's pretty likely that we'll see the first senolytics in hospitals in the next few years. Initially they'll be for specific conditions—IPF, arthritis, and so on—and only patients with symptoms will receive treatment. Gradually, if they prove safe and effective, we should look out for the more exciting prospect: trials of preventative senolytics, given to people with diseases at very early stages, or maybe no diseases at all, gradually encompassing broader and broader swathes of age-related ill health.

As well as pressing on with human trials, there's plenty more to be done in developing new senolytics. For example, D+Q kills about a third of senescent cells in mice. How much more powerful might its effects be if it killed 50 percent, or 80 percent of them? Also, different approaches work more or less well in different tissues: for example, dasatinib is better than another senolytic called navitoclax at killing fat precursor cells, while the drug-activated-suicide-gene method is better at clearing cells from the heart and kidneys than from the liver and intestines. Paradoxically, this could be good news—it shows that we don't need our first treatments to be perfect to have significant positive effects. Lifespan and health can be improved by a partial, imperfect intervention—with the chance of greater improvements as we get better at clearing these cells more comprehensively, across more tissues in the body.

The key question here is staring us in the face: if senescent cells are so damaging, and getting rid of them so good for mouse and human health, why do they exist at all? Why don't cells ticking any of the senescence-inducing boxes simply apoptose, vanishing without trace? The answer is that there are some places where senescent cells are more than just bad apples poisoning their local environment.

One example is during development. There are times when we're developing in the womb that evolution's solution to construct a particular structure in the body is to selectively kill cells. Sometimes this is done by apoptosis—most famously, our hands and feet grow as strange webbed paws and it is programmed cell death between the developing digits that separates our fingers and toes. In just the last few years, we've found that senescence is sometimes used instead. This may be down to the fact that development is a tightly choreographed process, and cells passing chemical messages among one another is key to its success: all those molecules pumped out by the SASP could be important, transient signals for nearby cells before the senescent cells are hoovered up by the immune system.

Once we're adults, senescent cells continue to play an important role besides cancer prevention: healing wounds. Say you get a cut in your skin: the injury sets an incredibly complex cascade of cellular and molecular action in motion. Nearby cells turn senescent and use their SASPs for good rather than evil: the pro-inflammatory molecules call for help from the immune system to clear up the mess and repel any opportunistic invaders trying to dive through the breach, and other chemicals in the SASP encourage tearing down the damaged structures and the growth of new ones to patch things up as fast as possible. (It's these pro-growth components of the SASP which can help to turn senescent cells' neighbors cancerous.)

Given these two examples, "senescent" is actually a poor choice of name for these cells: it implies that they're old, clapped out, useless, a feature of the end of both cellular and organismal life. It's entirely fair given how they were discovered, having hit a limit of growth after dividing too many times in a Petri dish. However, the SASP is a pro-growth, pro-healing balm in the right context, harking not from the end of life but from its beginning. It might even be that the late-life anti-cancer function of senescent cells is an afterthought, a case of evolution co-opting a cellular state whose original purpose was developmental to reduce the risk of cancer as we age.

There's a risk that removing senescent cells in some tissues could have unintended side effects. Some cell populations may be so small that senescent hangers-on are sub-optimal but nonetheless essential to maintaining function. One worry might be neurons: if a neuron is senescent but nonetheless an integral part of a memory, or a brain function you'd rather not compromise, it might be better to rescue it rather than finish it off.

There are a couple of other approaches to dealing with senescent cells which we could turn to in tissues where they're non-ideal but necessary. First, we could look for drugs which leave a cell in the senescent state but suppress the SASP, thus reducing the harm it does, which have been dubbed "senomorphics." Second, we could try to coax senescent cells back into the fold and turn them into normal cells again. This could be achieved by epigenetic reprogramming, which we'll discuss in a lot more detail in Chapter 8.

Eventually, the ideal outcome would be a visit to a gleaming clinic every six months or few years for a check-up, no more unusual than a trip to the dentist or optician. You'd have a few quick tests to establish the senescent cell burden in different parts of your body, and optimize your dosage of different drugs to target them in each organ; be prescribed a few pills, or given a few injections; then maybe stay in for a few hours to check that everything went okay, before heading home with instructions to take it easy, and a tube of lotion to speed the healing process in case you cut yourself in the next few days. We know from the fact that human lifespans in the rich world are pushing 80 years that senescent cells take decades to accumulate to really damaging levels, so infrequent treatment does seem plausible. In mice, the positive effects of senolytics lasted for months after a single dose—it's thus possible that humans may be able to go that long, or maybe longer between treatments.

For now at least, it seems hard to overstate the significance of the success of senolytics. Senescent cells are a fundamental driver of aging, present in many tissues in the body and implicated in a growing number of diverse diseases. The results in mice show that removing them can improve both how long and how well they live, with no obvious side effects. If we can get senolytics working in people, it will open up a cornucopia of new therapeutic options for diseases. It will also provide us with an irrefutable demonstration of the principle behind biogerontology: intervening in the aging process can pay dividends for life and health. No doctor would ever put "overburdened by senescent cells" on a death certificate. They are in a fundamentally different category from the diseases we fixate on in modern medicine—a cause rather than a consequence, related to many, maybe even most, of the diseases we worry about in old age. Quite apart from the drugs themselves, the idea of preventative senolytics—a universal, protective treatment for everyone, even the healthy—lays the conceptual groundwork for a medical revolution.

All of this means that a senolytic could well be the first true anti-aging potion to pass your lips.

REINVENTING RECYCLING: UPGRADING AUTOPHAGY

We've seen that a reduction in autophagy—the "self-eating" process of cellular recycling which clears up broken proteins, damaged mitochondria and more—is likely to be central to the aging process. Autophagy is our body's own way of getting rid of junk, so could we take advantage of this natural system to keep our cells pristine?

Given that dietary restriction (DR) increases levels of autophagy, eating substantially less could be one approach to activating the process and slowing our own aging. However, it would be even better if we could find some way to mimic the biological effects of restricting diet, but with less of the tedious abstinence. Enter "DR mimetics": drugs which activate many of the same mechanisms as DR itself (including autophagy) without the need to eat less.

The story of DR mimetics begins over half a

century ago, in November 1964, as the Canadian naval vessel **Cape Scott** left port in Canada bound for Easter Island. Easter Island is one of the most remote inhabited places on Earth, deep in the Pacific Ocean, some 3,500 kilometers from the coast of Chile. Named Rapa Nui in the native Polynesian, it's most famous for its monolithic **moai**: gigantic human-like stone statues with enormous heads. Prompted by plans from the Chilean government to build an international airport, intruding on centuries of almost complete isolation, the **Cape Scott** carried a 38-strong expedition team whose job it was to scientifically document the pristine island environment and its 949 native inhabitants before it was lost forever.

The expedition was eventful to the point of nearly failing entirely. At the time, Rapa Nui was a Chilean colony, and the arrival of the **Cape Scott** coincided with a chaotic but bloodless revolution: the islanders took Rapa Nui's only bulldozer hostage, Chile sent 40 marines to investigate the unrest and, at one point, the rebel leader Alfonso Rapu was forced to take refuge in the scientists' compound and escaped disguised in women's clothing. (Shortly afterward, the revolution succeeded, and he was elected mayor of the island.)

However, amid the chaos, meticulous scientific sampling of Rapa Nui and its inhabitants was taking place. Of 17,000 samples, medical records and X-rays collected, the most significant would turn out to be an unassuming vial of soil. The sample was returned to Canada and, four years later, given to a research group interested in finding new medicines in chemicals produced by bacteria.

The work was painstaking: bacteria were isolated from the soil, grown on plates of nutritious agar gel, then cultured with a range of test micro-organisms such as other bacteria and fungi to check for antibiotic effects.* One strain, **Streptomyces hygroscopicus,** was lethally effective when placed next to **Candida,** the yeast which causes thrush infections. The scientists isolated the chemical responsible for killing the fungus and dubbed it rapamycin, after Rapa Nui.

Rapamycin turned out to be much more than an anti-fungal agent. Further investigation showed that it was a powerful immune suppressant, and also stopped cells from multiplying. Though its effects on the immune system ended its short career as a potential fungal antibiotic, these two findings were far bigger: immune suppression is vital to stop patients from rejecting transplanted organs and its power to stop cell proliferation was immediately recognized

---

* This is a formalized version of the serendipitous experiment which gave rise to the discovery of penicillin, the first antibiotic, where blue-green fungus set up shop on a Petri dish which had accidentally been left open was found to have developed a halo around it in which bacteria wouldn't grow. The halo was caused by a chemical secreted by the mold which was toxic to bacteria—eventually isolated and named "penicillin" after the **Penicillium** fungus in which it was discovered.

as a potential new cancer treatment. After years of promising investigations, however, rapamycin still hadn't been formulated into a workable drug—and, out of the blue, pharma company Ayerst, which had been leading development, shuttered the program in 1982 as part of a restructuring exercise.

Scientist Suren Sehgal was dumbfounded. Convinced that this drug could be revolutionary, he sneaked a few vials of the rapamycin-producing bacteria home and hid them in his freezer, helpfully labeled "DON'T EAT!" There they stayed for five years, surviving a house move (with the freezer duct-taped shut and packed with dry ice), until he managed to convince his bosses to let him defrost the bugs and begin research on rapamycin once more.

You've probably realized by now that rapamycin isn't just an immune suppressant and anti-cancer drug (though Sehgal was right, and it is now licensed for use as both)—its biggest contribution to human health could be as anti-aging medicine. And if you one day take rapamycin or one of its derivatives to stave off old age, you have that ludicrous chain of events to thank. The Chilean government's decision to build an airport on Easter Island in the 1960s has already saved millions of lives—and, if rapamycin works against aging, it could rack up billions more.

Scientists trying to work out how rapamycin actually works discovered that it interacts with a protein which was named after the drug: "target of rapamycin," or TOR. TOR is a nexus in cellular metabolism, critical to some of the most funda-mental processes in life. The subtle variant of TOR found in humans and mice is known as mTOR, and both it and TOR work in basically the same way: they sense levels of sugars, amino acids, oxygen and insulin, and give instructions to other proteins in the cell based on what they find.

Rapamycin jams up one form of mTOR called mTORC1 such that it can't signal to the rest of the cell when food is plentiful. This effectively short-circuits nutrient detection, tricking the cell into thinking food is short, even if it isn't. At high doses, this can entirely stop cellular growth; at lower doses, rapamycin can dial down TOR, reducing growth and promoting autophagy.

As a result, rapamycin works an awful lot like DR and, like eating less, it extends lifespan in yeast, worms and flies. Based on both the common mech-anism and this evidence in simple model organisms, scientists set out in 2006 to do a rigorous test of ra-pamycin, along with several other anti-aging inter-ventions, in mice.

However, nothing involving rapamycin is ever simple. They intended to start treating the mice aged four months, but it took over a year to de-velop a method of coating the rapamycin so that it could survive both the food preparation process and the journey through the mouse stomach. By this time the mice were 20 months old—roughly equivalent to 60 in human years. It seemed likely that any effect would be dramatically reduced by starting this late: was there even any point in doing the experiment at all?

Their fears were misplaced, and the results of the study came as a surprise to everyone. They demonstrated something truly remarkable: rapamycin works, even when administered to mice which are already old. The mice had lifespans on average 10 percent longer than their untreated peers. This was a real breakthrough: not only the first ever demonstration that a drug could extend lifespan in mammals but, quite accidentally and even more impressively, a demonstration that it works even when started late in life. Subsequent studies have confirmed what will by now be a familiar refrain: that these longer-lived mice weren't struggling on in frailty, but staying younger for longer, with fewer and less severe age-related diseases. Rapamycin slows cell death and improves cognitive performance in the brains of mouse models* of Parkinson's and Alzheimer's, and improves the

---

* A "mouse model" is what scientists call mice genetically modified to be at risk of a human disease, either because waiting around for them to get it would take a long time or, in the case of some conditions like Alzheimer's and Parkinson's, because mice don't suffer from them at all. This means that there are caveats when trying to translate findings from mice to humans—but it's often a vital first step in understanding how new treatments work. Nonetheless, bear in mind both in this book and elsewhere that a mouse model is one step further removed from the clinic than an experiment in normal mice might be.

functioning of arteries in diabetic mice, probably by stimulating autophagy.

This is an impressive proof of principle for DR mimetics—after a tortuous tale taking us from a Pacific island via a home freezer to some unintentionally elderly mice, we have a result which sounds like it's crying out for human translation: even if started very late in life, rapamycin can extend lifespan and healthspan. Unfortunately, the problem with rapamycin is the side effects. Rapamycin is a laser-guided drug, targeting mTOR with precision—but mTOR is a target with immense tactical significance, a command center of the cell, and knocking it out has dramatic repercussions.

First, given that rapamycin's first use in patients was as an immunosuppressant, it will come as no surprise that it can suppress the immune system and increase the risk of infections—and there's no point aging more slowly if you just die of the flu instead. It also predisposes users to diabetes, making it a double-edged sword—postponing the likes of cancer and heart disease while bringing forward another of the harbingers of age-related ill health. Then there's the weirder stuff: hair loss, mouth ulcers, slowed wound healing, joint pain and, in male users, fertility problems (one study in mice noted that rapamycin reduced their testes in size by 80 percent!). While it's still a useful drug for transplant patients and cancer sufferers, people with no pre-existing conditions are far less likely to want to run that gauntlet for the chance of a modest reduction in the rate of aging.

However, treatments for our anti-aging arsenal may yet result from that expedition to Easter Island. First, those side effects are observed at far higher doses of rapamycin than are needed for its use as an anti-aging medication. In fact, some of them are even entirely reversed when it's given at lower doses—counterintuitively, while a high dose of rapamycin will suppress the immune system, a low dose doesn't suppress it a bit as you might expect, but seems to enhance its performance.

Furthermore, when pharmacologists get hold of a new drug, they often try tweaking it to alter its properties, and rapamycin is no exception: it is the parent of a family of derivatives called rapalogues. Experiments in mice with different rapalogues taken at different intervals and doses show that it is possible to maintain the benefits while reducing side effects.

Work on rapamycin and DR more generally has also inspired the search for other DR mimetics, some of which are undergoing clinical trials. One is metformin, a drug which has been in use since the 1950s as a treatment for diabetes. A trial of metformin in healthy older volunteers will begin shortly in the U.S.—and it's the first ever drug trial testing a drug's effects against the whole aging process rather than a particular disease. (We'll examine the scientific and wider implications of this groundbreaking trial in Chapter 11.) Though these efforts have been delayed by coronavirus, the scientists involved are trying to start a small-scale study to see if metformin could improve immunity in elderly people and thus strengthen their response against COVID-19.

There's also spermidine, which was first discovered (as you might guess from the name) in semen.* Spermidine activates autophagy, and has been shown to improve heart health and extend lifespan by 10 percent in mice, even if started late in life; suggestively, a study looking at the connection between diet and lifespan in humans found that those getting the most spermidine in their diet lived five years longer than those getting the least, even after correcting for other differences in their diets, lifestyle and general health. (Particularly high concentrations of spermidine are present in mushrooms, soybeans and cheddar cheese.) While observational studies should always be taken with a pinch of salt, together with the lifespan extension in mice this is exciting enough to inspire some proper trials, so watch this space.

Other naturally sourced contenders include resveratrol, a compound found in the skin of grapes; curcumin, one of the chemicals which makes turmeric yellow; aspirin, which on top of its many other

---

* Spermidine and the related compound spermine were first observed by pioneering Dutch microscopist Antonie van Leeuwenhoek when he noticed small crystals forming as he examined his own semen under his microscope in 1677.

physiological effects was recently found to enhance autophagy; and quercetin, which we met very recently as half of the D+Q duo. None of these quite has the firm evidence base to suggest that healthy people should take them preventatively, but there's plenty of biochemical diversity there for researchers to explore. As well as scouring the natural world, drug companies will also be searching for artificial compounds which either build on the capabilities of known molecules or come up with entirely novel ones. A company called resTORbio is trialing a new mTORC1 inhibitor called RTB101, which has shown promise in improving older people's response to a flu vaccine and reducing subsequent respiratory infections. The company is also trialing the drug for other age-related diseases, including Parkinson's, and, at the time of writing, a new study of RTB101 has just been announced to test whether it can reduce the severity of COVID-19 in nursing home residents.

Given that there are several at various stages from development to trials, and that many of these are natural compounds or existing drugs repurposed, DR mimetics are racing senolytics to be the first actual anti-aging treatment deployed in clinical care. (If metformin or RTB101 prove effective against coronavirus, they may win!) In common with senolytics, these drugs will probably first be used to treat a particular condition, whether that be COVID-19, or diseases where loss of autophagy is particularly relevant—neurodegenerative conditions seem a likely contender. If that works, patients taking the drugs will be under keen surveillance to see if there are any other, wider benefits to taking them. Ultimately, these medicines may become general-purpose preventative pills for multiple ills of old age.

DR mimetics all share the same broad strategy: tinker with known biochemical mechanisms to unlock the benefits of dietary restriction. The advantage is that DR is the most robust, long-standing anti-aging intervention we have; the downside is that we aren't expecting DR to have world-changing benefits in humans, so we wouldn't expect drugs which mimic it to have seismic effects either (though I'd definitely take a few years' extra healthspan if offered). The next step is to try to tinker with autophagy directly, rather than restricting diets or dealing with molecular middle managers in the cell. There are plans to engineer our own cellular recycling machinery which go beyond what our bodies can do on their own.

One problem which causes autophagy to falter with age is that, over time, the system can literally get clogged up with gunk. Autophagy takes place in compartments in the cell called lysosomes, which are like tiny, itinerant stomachs, floating around ready to digest the waste which is ferried to them by various different cellular refuse collectors. Like the stomach, the lysosome's interior is acidic and full of digestive enzymes, each specialized in chopping or smashing or ripping apart particular types of molecular waste product.

Unfortunately, there are some kinds of waste so mangled that none of the sixty or so enzymes in the lysosome can work out how to crowbar it open. This isn't such a problem at first—if you've got some undegradable troublemaker floating around in the cell, you could do worse than incarcerate it within the lysosome, away from all the important, delicate cellular components. However, there comes a point where the lysosome is so bloated with rubbish that it's no longer able to work at maximum efficiency.

The rubbish is known as "lipofuscin" and it's made up of broken and misfolded proteins and fats, crosslinked together along with highly reactive metals like iron and copper. It's easy to spot under a microscope because it's fluorescent—if you shine light of a particular color on it, it glows back at you in a different color. Lipofuscin is a particular problem in non-dividing cells, like those in the brain and the heart—this is one reason neurodegenerative conditions may be the first targets of autophagy-enhancing drugs. Cells which are constantly replicating can sidestep the buildup by dividing waste between each daughter, and a problem shared is a problem halved—perhaps even a problem solved if there is some threshold level above which lipofuscin starts to become an issue, and continual halving can keep levels below that threshold. This could happen if, for example, too much lipofuscin dilutes the acid inside the lysosome to a point where some of its enzymes, which require an acidic environment to do their job, stop working. This in turn would cause further garbage to accumulate—garbage the cell was previously able to break down—and set a vicious cycle in motion. This idea is known, delightfully, as the garbage catastrophe theory of aging.

One place where this vicious cycle is particularly problematic is the eye, in a condition known as age-related macular degeneration, or AMD. AMD is the most common form of blindness in rich countries,* and the majority of people over the age of 80 have at least some signs of it. The disease is caused when retinal pigment epithelial (RPE) cells die. These cells support the light-sensitive rods and cones on the back of the eye, and their death causes loss of vision in the macula—the central region of the eye responsible for high-resolution color vision.

One of the prime suspects for this cell death is lipofuscin: in old age, distended lysosomes stuffed so full of a vision-related waste product called A2E can take up fully 20 percent of the volume inside the cell. And it's not just our eyes where lipofuscin seems to be behind age-related problems: another example of lysosomes stuffed to bursting is found in the

* Globally, AMD is substantially outstripped as a cause of blindness by cataracts—which can usually be cured by a simple operation—and, most distressingly, "uncorrected refractive error": people who need glasses, but don't have them.

foam cells which make up atherosclerotic plaques, which are made up of immune cells with lysosomes clogged with cholesterol, particularly in oxidized and glycated forms which are hard to digest.

These lysosomal vicious cycles might be worth trying to break. One suggestion is inspired by how we currently treat lysosomal storage disorders (or LSDs), a collection of rare conditions resulting from mutations in the genes coding for the various enzymes the lysosome contains. If one of those sixty junk-zapping enzymes is broken or missing, sufferers are left unable to break down the particular waste product which that enzyme degrades, causing their lysosomes to fill up with it at an unfortunately rapid rate. Not coincidentally, there are about sixty LSDs, with the worst types causing death in infancy. However, some of them can be effectively cured by providing patients with the enzyme they're missing, allowing many patients (depending on the LSD) to lead relatively normal lives.

In aging, rather than providing a replacement for an enzyme which is missing from the body, we'd need to provide new enzymes which can help lysosomes deal with garbage they can't currently tackle. One source could be bacteria. Bacteria are incredibly versatile organisms and have worked their way into almost every conceivable ecosystem on the planet, finding ways to subsist on incredibly unlikely foodstuffs. Thus, it seems pretty likely that, for any given type of lipofuscin, there's a bacterium somewhere which can make a living digesting it. Several different groups of scientists are taking this route, trying to narrow the search for enzymes which could break down products which our cells alone cannot.

Multiple different species of bacteria have been identified which can degrade the cholesterol-based waste products which are a problem in atherosclerosis, from habitats as diverse as North Sea sediment and piles of manure. One particularly fascinating study discovered some candidate cholesterol-crunching enzymes in **Mycobacterium tuberculosis,** the bacterium which causes tuberculosis infections in humans. Tuberculosis is an infection with incredible skills when it comes to evading the human immune system: its bacteria are capable of being engulfed by and then **hiding inside** the macrophages sent to dispatch them. How they survived inside these immune cells was a long-standing mystery, until it was discovered that they were able to subsist by extracting energy from none other than cholesterol inside the cell. When the genes responsible were transferred into human cells in a dish, the cells gained the ability to break down cholesterol. Unfortunately, further work is needed here because the products the cholesterol was broken down into turned out to be toxic. But it would be a neat symmetry if tuberculosis—probably one of the leading killers in human history—helps us create tools to defeat cardiovascular disease, one of humanity's deadliest foes today.

Similar work on A2E, the lipofuscin associated with age-related sight loss, has identified enzymes which can break it down, too. The furthest advanced in development is called manganese peroxidase, usually found in fungi which live on dead wood, which use the enzyme to break down lignins—tough materials which give strength to wood and bark. A 2018 paper by a startup called Ichor Therapeutics got as far as injecting a modified version of it into the eyes of mice, and demonstrated that it rapidly cleared out both A2E and a number of other by-products of visual chemistry which accumulate in lysosomes of RPE cells. They are hoping to turn this preliminary work into a treatment called Lysoclear.

Once suitable enzymes have been isolated and optimized for both efficiency and safety in humans, we should be able to inject them just as we do to treat LSDs, or use gene therapy (more on which in Chapter 8) to allow our cells to manufacture these enzymes themselves.

A final approach is to convince cells to take out the trash, rather than hoarding it in their lysosomes. This could be advantageous if lipofuscin is made up of too many diverse products for a small number of additional enzymes to digest. Once outside the cell, the best-case scenario is that macrophages swing by and pick up the debris, before getting rid of it entirely. There's already a drug called Remofuscin which seems to do exactly this for A2E in the retinal cells of both mice and monkeys, and is currently undergoing trials for Stargardt disease, a genetic condition in which macular degeneration is accelerated to the point where it affects children rather than the elderly. If it works, its use could be extended to the age-related version of the disease, and it or similar drugs could convince other lipofuscin-stuffed cells to get rid of their toxic stockpiles.

All in all, there are quite a few options available to us to improve our bodies' built-in recycling capabilities, by either convincing our cells to do more of it or augmenting its capabilities to get rid of otherwise unrecyclable waste. Drugs that mimic the effect of dietary restriction and treatments to help our cells deal with accumulating garbage are in development, and may save our sight, our minds and more from deterioration with age.

AMYLOID

We've seen that one nasty property of some proteins is that they aggregate into amyloids. These are proteins which, misfolded in particular unfortunate ways, gain the ability to stick together into clumps. Where the normal version of a protein will happily putter around doing its thing and minding its own business, an amyloid-prone misfolded version will seek out others like itself and latch on, in a kind of protein clone conga. Individual strings of amyloid are known as fibrils and they can aggregate into larger structures called plaques.

Plaques are most commonly associated with Alzheimer's disease—they were first observed by Alois Alzheimer, who found strange plaques between the cells and "tangles" inside them in the brain of a patient who had died at 55 with the dementia which now bears the doctor's name. It was 80 years before we had the biochemical and genetic tools to work out what these strange substances were, and to formulate a coherent theory for what caused the disease.

The first solid clue came from cases of "early-onset" Alzheimer's: patients who tragically come down with dementia as early as their twenties—though more commonly in their forties and fifties, rather like Alzheimer's first patient. Dementia is normally very starkly a disease of aging: it's almost unheard of before the age of 60, but then its occurrence rises even faster than risk of death, with the chance of an Alzheimer's diagnosis doubling every five rather than eight years. These early-onset cases are therefore remarkable: why do these patients get dementia decades before everyone else?

Years of genetic detective work eventually narrowed the problem down to mutations in a single gene: **APP**, the gene that codes for amyloid precursor protein, or APP. APP is normally chopped into three fragments, each of which has different jobs in the brain. About 10 percent of the time one of these chops will result in a fragment of "amyloid beta" being produced. This is happening all the time, in all of us, but with age something changes which either increases the production or reduces the clearance of amyloid beta (or both), and enough of it lingers that it sticks together to form plaques.

The mutated forms of APP which drive early-onset dementia are more likely to produce amyloid beta, so people carrying them build up plaques far more rapidly than people with the normal version of the gene. The appearance of amyloid plaques correlates with the onset of the disease and provides evidence that too much amyloid is enough to cause Alzheimer's.

However, decades of further investigation and failed therapies have served to undermine this "amyloid hypothesis." People without dementia often have extensive amyloid plaques, and those with the disease can be surprisingly amyloid-free. Parts of the brain with the most amyloid often fail to correspond with the parts of the brain thought to be worst affected based on patients' cognitive symptoms. The biggest challenge to the amyloid hypothesis is the graveyard of unsuccessful drug candidates—every one of dozens of attempts so far to interfere with the creation of amyloid or clear it up after it's produced has failed to have any effect on the symptoms of dementia sufferers. This isn't because we can't clear the amyloid: the latest immunotherapies have shown significant success by using antibodies, molecules which stick to the plaques and encourage the immune system to clear them up, to rid the brain of amyloid. Brain scans of patients who have had these treatments seem to be almost completely clear of the misfolded protein . . . they just don't see any actual functional improvement, rendering this impressive technical achievement somewhat moot.

The amyloid hypothesis staggers on anyway. What might be the final theory in its defense argues that we have simply intervened too late in previous clinical trials: we need to catch the amyloid before it destroys neurons and sets the other dominoes, perhaps some yet unknown, toppling. Thus, new trials are addressing this timing question in patients with early-onset Alzheimer's. Genetic testing for the **APP** mutation (and a few others) allows patients to be identified many years before an almost certain prognosis, meaning that they can be started on immunotherapy well before the disease begins. In the next few years, we should get the first results of these trials.

There are other possible culprits in Alzheimer's. Those tangles which Alois Alzheimer found inside brain cells are aggregates of another protein called tau, and therapies to slow its production or remove it are in the works. Dementia also seems to correlate with diabetes, leading scientists to wonder whether the way the brain deals with sugar and insulin could be a key component of the disease. Some theories suggest that Alzheimer's is driven by infections, with suspicion aroused by finding herpesviruses and bacteria responsible for gum disease entombed in the plaques in sufferers' brains. Others suggest that dementia is driven by inflammation, and that anti-inflammatory immune-calming strategies might help—which, given the prominence of inflammation in aging, is highly plausible. Scientists are also investigating whether the systems which allow waste products to drain from the brain may be impaired as we get older. Sleep seems to be important, with the brain using its downtime to flush out waste products, including amyloid; sleep duration and quality are known to decline with age, so this could be another factor. It's quite likely that several of these theories will prove to be partly responsible and uncovering their interplay may be crucial for a total cure.

However, all of these alternative theories still need to explain early-onset Alzheimer's, in which beta amyloid aggregation alone seems to set off cognitive decline. We know that beta amyloid aggregation is something which happens in many of us—even in people without symptoms of dementia, 20 percent have detectable beta amyloid by the age of 65, and almost half have some by 90. Maybe, given long enough, beta amyloid would get us all. Thus, it might be a reasonable precaution to clear aggregating amyloids out of everyone's brain. At a minimum, it's good to have these reliable anti-amyloid treatments if only because we know that young and healthy brains don't have these deposits, while old and diseased ones do. Perhaps amyloid-clearing drugs will be needed together with tau-busting, anti-inflammatory or entirely left-field future Alzheimer's treatments.

Though Alzheimer's is the most famous place where amyloids play a role, we are continuing to learn how similar aggregates of different proteins play a part in many other diseases. We've already met alpha-synuclein, which forms amyloids in Parkinson's; other neurodegenerative diseases like ALS (which causes death of "motor neurons" that control muscles) and Huntington's have aggregates of their own, formed from different misfolded proteins. Type 2 diabetes is accompanied by amyloids made from a protein called (confusingly) amylin. Though many amyloid diseases are caused or substantially worsened by genetic mutations which cause the proteins to clump together, some aggregation happens to all of us as a part of normal aging.

One amyloid which gets far less attention than amyloid beta is formed from transthyretin, or TTR. TTR is a blood-borne protein which ferries thyroid hormones and vitamin A around the body. It's evidently a protein right on the cusp of amyloid formation, because there are over one hundred known mutations which cause it to form amyloid: any one of many tiny changes can push it over the edge into rapid aggregation. TTR amyloid can build up all over the body in older people, earning the condition the name senile systemic amyloidosis, or SSA. It's often worst in the blood vessels, where amyloid's effects on the cells lining vessels can narrow and stiffen them, and the heart, where it strangles the muscle and disrupts the electrical signals which cause it to beat. Ultimately, this leads to heart failure, a diagnosis meaning that the heart can no longer pump enough blood to adequately supply the whole body. This is a common condition in elderly people, with a variety of different causes, and it's thought that TTR amyloid may be an underappreciated one.

The problem is that cardiac amyloidosis is hard to diagnose. No one wants to take a heart biopsy, let alone of an old patient with heart problems, and the non-invasive tests for it are somewhat specialist (they include MRI and PET scans, which are not normal diagnostic tests when an older person turns up in the hospital with heart problems). The other issue is cultural: autopsies are very rarely performed when an old person dies. If an 82-year-old finally succumbs to heart failure, probably alongside a few other medical conditions, no one is likely to call in a pathologist to carefully dissect the body and determine exactly what killed them. This is another symptom of our medical and scientific practice being biased against aging: "dying of old age" is considered unremarkable, and rarely merits detailed investigation. It's not that we need the precise cause of death to be determined for every older person who dies, but it would be useful to have some more data about what pathologies are present in order to work out what biogerontologists should be trying to fix most urgently, and what problems might be nascent in elderly people, waiting to strike us down if we just lived a bit longer.

Amyloidosis is definitely a candidate killer-in-waiting. A study in Finland found that, on autopsy, 25 percent of over-85s had TTR amyloid on their hearts—rising to more than half of those who died aged over 100. Another in a Spanish hospital found that 13 percent of patients with one type of heart failure had significant amyloid deposits, something which would almost certainly have gone undiagnosed were it not for their research. Those at most risk are probably the very oldest old, who have had the good fortune to dodge all the other age-related diseases—it may well be the slow accumulation of TTR amyloid that kills them. SSA is hypothesized to be the leading cause of death in supercentenarians, people who live more than 110 years.

Like the amyloid beta responsible for Alzheimer's, TTR and other amyloids are in researchers' crosshairs. There are immune therapies, like the antibodies which have been successfully used to clear amyloid beta from the brains of Alzheimer's sufferers, in development for TTR amyloid as well—one example is a drug candidate named PRX004, under development by a company called Prothena.

Also under investigation are catabodies, a type of antibody which, instead of labelling something as a target for the immune system, destroys it directly. We actually produce catabodies naturally—against amyloid beta, tau and TTR at least, and maybe against more misfolded proteins. However, our natural defenses are insufficient and, after identifying some appropriate catabodies to optimize, this idea is being taken forward as a treatment called Cardizyme by another company, Covalent Biosciences. They are also developing Alzyme and Tauzyme to target the amyloid beta and tau aggregates in Alzheimer's. All three have been shown to clear their respective amyloids in mice. Catabodies have two key advantages over antibodies: first, because they don't just cling on to a target until the immune system destroys it but smash it themselves and move on, the same catabody can work over and over again on many molecules in an aggregate; and second, because they don't call the immune system's attention to the problem, they create far less inflammation, which, as you know by now, is best avoided where possible.

There's also a potentially exciting approach which uses a chemical commonality which is shared by a number of different kinds of amyloid to break them all down with a single therapy, with a rather incredible backstory. Known as "general amyloid interaction motif," or GAIM, it was discovered entirely by chance in a bacteriophage (or phage for short), a type of virus which doesn't infect humans, but bacteria. The virus, M13, was first discovered in the sewers of Munich in 1963. It went on to become a staple of lab biology and, in the early 2000s, Israeli scientist Beka Solomon used M13 phages in an attempt to ferry more of an anti-amyloid antibody she was developing into the brains of mice with Alzheimer's disease. To her surprise, the control group of mice which received only the virus and none of the antibody showed significant cognitive improvement, a finding which made no sense, because M13 infects **E. coli** bacteria, and shouldn't have any effect on human cells or proteins.

It turned out that, by an incredible coincidence, the molecular lock and key which M13 uses to gain access to **E. coli** cells is remarkably similar to a molecular structure found in many kinds of human protein aggregates. As well as granting access to bacterial cells, the viral protein could break apart aggregates of amyloid beta and tau in Alzheimer's, the alpha-synuclein in Parkinson's, the huntingtin behind Huntington's disease, and even the aggregates behind motor neurone disease and CJD (a rare human brain condition made famous by the "mad cow disease" crisis in the 1990s). It's hard to overemphasize how bizarre this is, but nonetheless GAIM has been shown to clear both amyloid beta and tau in mouse models of Alzheimer's disease, and to improve their cognitive function. Human trials are ongoing, headed up by a company called Proclara Biosciences.

Ideally, one or more of these treatments will be turned into preventative therapies for amyloid-based diseases. Perhaps we could all get injections of anti-plaque drugs at regular intervals to prevent

the buildup of these toxic aggregates or, even better, be immunized against many different amyloids
in childhood along with measles and diphtheria.
Intervening before we're old and diseases are too
far advanced is exactly what the treatment of aging
should be about, and all of these therapies have the
potential to be used preventatively. However we go
about it, clearing amyloids is likely to be important
to treating aging.

In this chapter, we've explored how to remove
senescent cells and problem proteins. In the next,
we'll look at places where simply getting rid of things
is not a sufficient treatment, and we're going to need
to replace and rebuild them in our aging bodies.

6

In with the New

There are some aspects of our aging biology for which getting rid of the bad actors won't be a sufficient therapy unless we can replace them with something better. For example, though the aged immune system can be dysfunctional and put us at greater risk of infectious diseases and cancer, it's better than the alternative: the only thing worse than having an aging immune system is having no immune system at all.

Thus, we need to come up with ways to reinforce our ailing defenses, and far more besides, providing our biology with a helping hand to reverse some of the decline of aging. This chapter focuses on four broad categories of replacement therapy: first, stem cell therapies, where providing stem cells can underpin the regeneration of many different parts of our body; then, the immune system, where various ideas (including some stem cell therapies) could help restore it to a more youthful state; next, the good guys in our microbiome—the huge ecosystem of bacteria, viruses and fungi which we carry in our guts, on our skin and elsewhere—which may also need topping up with age; and finally, the long-lived scaffold of proteins outside our cells which suffer chemical damage over time, where replacement may be a more promising approach than repair.

STEM CELL THERAPY

Stem cell therapy is one of the hottest areas in medicine, and harnessing stem cells for treatments is very likely to be a key weapon in our arsenal against aging. Stem cells will help to replenish cells that are lost during the aging process, playing a role in diseases from age-related blindness to diabetes and Parkinson's.

However, given the hype which often surrounds it, stem cell therapy is frequently misunderstood. The term "stem cell" is thrown around by charlatans who will take desperate patients to shiny, semi-regulated clinics and infuse them with mysterious solutions to "cure" all kinds of different ailments. "Stem cells" aren't a single thing, nor are they some kind of elixir where a single treatment will fix many different diseases or undo the ravages of time systemically. In order to understand the huge, genuine potential of these therapies, we need to understand exactly what stem cells are, and therefore what we can expect them to do. Getting the right cells to the right place at the right time is key to using them for regenerative medicine.

The definition of a stem cell is a cell which has a choice when it divides: it can do what most cells do when they divide, and form two of the same kind of cell (which in this case would be two stem cells, used for replenishing the stem cell population); it can divide into a stem cell and another kind of cell (thus not depleting the stem cell population, and adding a fresh, new cell to wherever it finds itself, from the skin to the lining of the intestines); or it can turn into two non-stem cells (to maximize tissue replenishment at the expense of the stem cell population). The process of turning from a stem cell to a specific type of body cell is "differentiation." It's easiest to think about these abilities in the context of a developing embryo.

We all start out as a single fertilized egg, which is the grand matriarch atop a sprawling family tree of different cell types. She is the ultimate generalist, able to form every cell in a developing baby. Her first few daughter cells in the very early embryo are "pluripotent" because of their ability to form any tissue of an adult human.* Pluripotency is fleeting: it's not long before all the cells in the developing embryo are merely "multipotent," still retaining pretty eclectic career options but no longer able to become literally anything. As development continues, so cells' potential fates narrow as their position in the body becomes clearer. A cell might start out as a pluripotent stem cell, and some of its daughters would go on to be general-purpose brain precursor cells, and some of **their** daughters highly specific kinds of neurons with a particular role in the brain.

Eventually, most cells make it to the end of the road—they are said to have "terminally differentiated." This means if you're, say, a particular type of cell in the heart or liver, that's your job for life. If you divide, your daughters will be two of the same kind of cell you already are. The handful of cells which hang back from terminal differentiation are "adult stem cells," like the populations of cells which maintain our skin, the lining of our intestines, or the hematopoietic stem cells (HSCs) which produce hundreds of billions of fresh blood cells on a daily basis.

---

* Technically, the very first few daughters are said to be "totipotent" and have greater powers still—not only can they build any cell of the body, but also any cell type outside the embryo which forms part of the interface between mother and developing baby, like the embryo's parts of the placenta. Pluripotent stem cells can form any body cell, but only a minor part of the placenta known, catchily, as the "extraembryonic endoderm."

This brings us to the first category of stem cell therapy: the transplantation of adult stem cells from one individual to another, or from the same individual to themselves. Though stem cell therapies are often thought of as quite futuristic, there is one workaday stem cell treatment which we've been doing successfully for half a century. Bone marrow transplants, which are more properly known as HSC transplants because the HSCs are often sourced not from bone marrow but other locations such as a donor's blood or an umbilical cord, are now a routine (if serious) medical procedure.

The classic scenario is treating a blood cancer like leukemia. In leukemia, the body massively overproduces particular kinds of blood cells, which fill up the bone marrow and overwhelm the stem cells there. This means that they lose the ability to produce blood cells, and patients most commonly die of infections due to a lack of the immune white blood cells which fight them. Like most cancers, the standard treatment is chemotherapy or radiotherapy, both of which preferentially kill the fast-dividing cancer cells, hopefully not doing too much damage to other fast-dividing cells in the process. However, HSCs are incredibly sensitive to these treatments, too, and they can be depleted so badly that you'd end up dying anyway, just of a catastrophic loss of blood cells instead of cancer. The solution is to wait until the therapy is finished, then inject some HSCs to resume production of blood cells.

There have now been well over one million HSC transplants globally and tens of thousands more are performed every year—they are an incredibly successful procedure which has saved huge numbers of lives. However, using adult stem cells has limitations, especially when it comes to treating aging. One key problem is that you can only do it where a suitable population of stem cells exists—there don't seem to be heart stem cells, or stem cells for most parts of the brain, for example, though scientists continue to search. Even if there were useful brain or heart stem cells, most of us would be understandably hesitant to sign up as donors: in the vast majority of cases, HSC donation from bone marrow requires taking drugs for a few days, followed by a few hours having your blood filtered to extract the stem cells, which isn't too onerous; extracting heart or brain cells, by contrast, could be quite a risky and invasive procedure for the donor.

The second problem is immune rejection. Just like in an organ transplant, the recipient patient's immune system may identify the new cells as "nonself" and destroy them, removing the benefit of the treatment and mobilizing an immune overreaction which can, in the worst cases, lead to death.* Just

* In the case of HSCs, the reverse can also happen: since some of their daughters are immune cells, this donated immune system can recognize its new home as non-self, and go on a dangerous rampage, "graft-versus-host disease." Doctors actually turn this effect to their

over half of HSC transplants use the patient's own cells, which avoids this problem, and we've got pretty good at matching donor to recipient for HSCs—but even a decent match still means that, like people who receive organ transplants, many recipients face a lifetime of immune-suppressing drugs, which can have serious side effects and put them at risk of infection.

The breakthrough which could solve these problems arrived in 2006, when Japanese scientist Shinya Yamanaka managed, for the first time, to turn back the developmental clock in adult cells and revert them to a pluripotent state—meaning they had the potential to become any kind of cell in the body. The ultimate medical hope here is that we could produce unlimited quantities of cells of any type from a patient's own cells, without the need for potentially invasive donation procedures, or even for relevant stem cells to exist at all. And, since we could generate these cells from the patient, there would be no risk of immune rejection either.

It was thought for a long time that the process of development and differentiation was entirely unidirectional, from fertilized egg, via pluripotent and multipotent stem cells, to adult cells in the body.

---

advantage in the case of diseases like leukemia where the donated immune system will target any remnants of the cancer that survived, which is "graft-versus-leukemia effect." This is considered just as important a part of the treatment as repopulating the HSCs.

Perhaps with hindsight it should have been obvious that it isn't: after all, the miracle of pregnancy requires two adult cells—an egg and a sperm—to merge and, in doing so, turn the clock back from being hyper-specialized reproductive cells to being a fertilized egg once again, with a reawakened ability to become any cell in the human body. Dedifferentiation, then, wasn't against the laws of biology; the question was, then, could the process be reproduced in the lab?

In a series of pioneering experiments in the 1960s, British scientist John Gurdon showed that we can. He took the nucleus—the part of a cell containing the DNA code—from a frog cell, placed it into a frog egg cell whose own nucleus had been destroyed, and watched what happened. Nuclei from young embryos transferred into an egg cell could become adult frogs, while nuclei from adult frogs didn't make it as far, often failing entirely but sometimes making it as far as a late embryo with distinguishable body parts.

This technique of transferring an adult cell nucleus into a vacant egg cell has been refined over the years and made more reliable. In 1997, it was responsible for the conception of probably the world's most famous sheep: Dolly, the first ever mammal to be cloned. The transferred nucleus meant that she shared exactly the same DNA as her "mother" whose body it had been taken from.

Clearly a fertilized egg contains some kind of

machinery which can "reset" the changes which cause cells to differentiate. By the 2000s, Yamanaka's lab was studying the genes at work in embryonic stem cells, or ESCs—cells extracted from embryos sufficiently early in development that they're still pluripotent—to find a way to emulate whatever heady chemistry in an egg cell allows it to turn back the clock. He and his lab eventually succeeded, identifying four genes, the "Yamanaka factors," that, when transferred into a cell, could induce pluripotency. This feat—the creation of "induced pluripotent stem cells," or iPSCs—won Yamanaka a Nobel Prize, shared with Gurdon, in 2012.

The reason that the ability to turn back the differentiation clock is worthy of a Nobel Prize isn't just for the pluripotent cells themselves, it's for what they can be used to make—which seems to be literally any kind of cell. As proof, scientists have tried swapping out the embryonic cells in a very young mouse embryo for iPSCs—and the result, eventually, is adult mice with every cell type working as they should. This shows that, given the right environment (in this case, inside a mouse embryo), these cells can be coaxed into becoming any cell type in an adult mouse.

However, this is perhaps another cheat—putting iPSCs into nature's own ready-made cauldron and creating an adult mouse is one thing, but what we really need to do is to produce cells of a given type on demand. So, having dedifferentiated a cell to become an iPSC, the challenge is then reversed—how do we redifferentiate it again into the cells we want? The answers are found by looking back at how embryos develop: if we can understand how the cells in a growing mouse or human "know" what to become, we can simulate those conditions in a dish in the lab, and produce whatever type of cell a given patient needs.

A cell in a developing embryo knows what it should become thanks to a constant stream of chemical messages coming from cells, near and far. Developing cells secrete many different molecules, and the strength, timing and duration of these chemical signals are determined by the signals they receive. This recursive, decentralized system gives rise to patterns upon patterns of different chemical messages, allowing each cell to work out where it is by the chemicals in its local environment, and therefore what it should grow up to do.

So, if you want to encourage an iPSC to turn into a neuron, a heart cell, a skin cell, or whatever, you need to supply it with the appropriate series of signals—the same succession it would receive if it were developing for real in a complete embryo, rather than sitting in a dish in a lab somewhere. Over days or weeks, scientists drip the relevant signaling molecules onto the cells in question, slowly guiding them to their required fate. Our improving understanding of both embryology and cell culture means that scientists are getting increasingly good at making the cells they want in the lab.

Hopefully it's obvious why this would be great news for cell therapies, and not just against aging: anywhere cells are lost, due to disease, injury or the aging process, we could manufacture new ones to take their place. And, in the ideal case, cells could be derived from the patients themselves, meaning that there will be no struggles to identify a matching donor—the immune system will happily recognize them as "self" and not flare up in an attempt to rid the body of the invaders.

While it might sound like pluripotent cells are somehow the "best" stem cells, given their ability to form any kind of cell, it's worth emphasizing that stem cell therapies won't be injecting the iPSCs themselves. This is because iPSCs are no real use in the body because, in the absence of guiding signals, they won't turn into the type of cells needed—and, not only that, but they can pose a risk of cancer. Pluripotent cells can reproduce indefinitely in a dish in the lab, which is great if you want to make a huge batch of them for an experiment, or to replace lost cells in the body. The flip side of this is, if any pluripotent cells remain in the mixture you inject into the patient, they have the potential to divide indefinitely there too, forming a tumor.

The particular type of tumor caused by pluripotent cells is a teratoma, and they're absolutely grotesque. In fact, they're so horrible that I highly recommend seeking one out in person to truly fathom their monstrousness (even the name "teratoma" derives from the Greek for "monstrous tumor"). Teratomas can occur naturally, though mercifully rarely, usually in women's ovaries or the testes in men. Without those carefully choreographed signals used in development, the pluripotent cells don't really know what to do—instead, they differentiate almost at random, forming a chaotic lump of horror. This combination of rarity and ickiness made them collectors' items for Victorian medics—you can find them in anatomical museums, floating in formaldehyde, disgusting balls of muscle, matted hair, teeth, bone, fat and sometimes even eyes and bits of brain. One particularly unpleasant example, extracted from the ovary of a patient in Japan in the early 2000s, looked like a tiny, misfolded baby, complete with hair, proto-limbs, a few teeth and a single malformed attempt at an eye. It's therefore critically important that any pluripotent hangers-on are removed before delivering your stem cell therapy to a patient.

There are many different ideas for therapies using differentiated daughters of iPSCs for therapy, and many of the therapies leading the way are for diseases of aging. The ideal test case for a cell therapy is one where a disease or dysfunction can be attributed to loss of a single type of cell, meaning that you need to replenish only that cell type, and not a complex population. As a result, two of the fastest-moving therapies are to replace RPE cells, which we met in the previous chapter, to alleviate age-related macular degeneration, or AMD, and the specific type of neuron whose loss causes Parkinson's disease.

The stem cell treatments which are closest to realization are probably those for AMD. Two trials in 2018 used stem cells to make RPE cells and implant them into patients' eyes. Both were "Phase I" trials designed to check that a treatment is safe rather than prove whether or not it works—but not only did they demonstrate safety, both also showed improvements in participants' vision. One patient went from reading at an infuriating 1.7 words per minute to a passable 50 after the treatment, and was able to read 29 more letters on one of those visual acuity test charts where the letters gradually decrease in size. These studies involved a combined total of just six patients, so much more work is needed, but these are exciting preliminary results.

The most significant shortcoming with these trials is that they used **embryonic** stem cells to make the RPE cells, and so by definition they can't have come from the patients being treated. The patients therefore needed to take immunosuppressive drugs to stop their immune systems from attacking the new cells. The next step is to replicate these positive results with patient-derived iPSCs. The first test in humans took place in 2014 in Japan but the trial was stopped for safety reasons after the discovery of potentially cancerous mutations in the implanted cells. Though the patient didn't experience any problems, it caused scientists to slow down and take stock. A 2019 study by the U.S. National Eye Institute sought to allay any fears, using a painstakingly realistic manufacturing protocol for RPE cells, carefully checked at every stage for safety. The protocol passed this meticulous testing, and the next step is to try it in human patients.

The success of ESC-based treatments plus the positive steps toward use of iPSCs mean that hopes are high for patients' own cells being used to treat age-related vision loss in a hospital near you in the not too distant future, representing the first clinical triumph to stem from Yamanaka's 2006 discovery.

Parkinson's is caused by the loss of "dopaminergic" neurons—specialist neurons that produce a chemical called dopamine that brain cells use to communicate with one another. By the time symptoms appear, patients have lost up to 80 percent of these neurons, massively undermining the precise systems in our brains which allow us fine-grained control of our movements. The standard treatment for advanced Parkinson's patients is L-dopa, a chemical which the brain can turn into dopamine, but the appeal of stem cell treatments is hopefully obvious: they could replace the dopaminergic neurons and potentially cure the disease, rather than masking its symptoms.

Stem cell treatments for Parkinson's disease have a surprisingly long history. The first pioneering operations took place over three decades ago, in Lund,

Sweden, in 1987. They involved surgically graft-
ing dopaminergic neuron precursors taken from
aborted foetuses into the brains of two patients
with advanced Parkinson's disease. The idea was
that these immature cells would multiply and develop into dopaminergic neurons—and it seemed to
work. Buoyed by this early success, the experimental
operations carried on for many years. The results
were incredibly compelling: one subject, Patient 4,
received his graft in 1989 and improved so dramatically over three years that he no longer needed to
take L-dopa. He enjoyed near-complete remission
for nine years, when deteriorating motor function
meant that his medication needed to be gradually
started again. When he died, 24 years after the operation, a post-mortem revealed that the transplanted
neurons were still alive, and had made connections
to the surrounding brain cells—though any functional benefit was lost by this point, probably as dementia and general deterioration took hold in the
rest of his brain.

After these promising early results, the story took
a number of twists and turns: the Swedish studies
involved only 18 patients in total, and larger, more
thorough trials conducted by the National Institutes
of Health in the U.S. seemed to cast doubt on the
effectiveness of the treatments; the Swedes hit back,
saying that the bigger projects hadn't used fresh cells
in their transplants, failed to use immunosuppressant drugs to stop the graft being rejected and didn't

watch their patients for long enough to observe an effect (Patient 4's improvements, remember, had taken three years to become apparent). There's little enough data that debate continues to rage, but the clearest sign that the neurology community retained its excitement and optimism is that, in 2010, a collaborative effort to do the definitive study on fetal stem cells for Parkinson's began. Involving more than 100 patients across Europe, the first results from the TRANSEURO study are expected in 2021.

Unfortunately, since they can only be extracted from aborted foetuses at a specific stage in development (the painstaking procedure involves finding a region the size of a pinhead in a foetus just a couple of centimeters long), the supply of fetal dopaminergic neuron precursors would severely limit the availability of this treatment. As a result, if further work shows that these cells can help with Parkinson's, the obvious next step is to use iPSCs to make them instead. The first trial with iPSCs—albeit not from the patients themselves, but other donors—began in Japan in 2018, and more are expected soon.

Though AMD and Parkinson's treatments are leading the pack, there are plenty of other stem cell therapies close behind them. Diabetes might be next: we can make beta cells—the insulin-producing cells found in the pancreas that keep blood sugar under control—from iPSCs in the lab, and they can cure mice with diabetes. Human iPSCs have also been used to make "chondrocytes" which make and repair the cartilage found in joints, and they have been successfully used to regenerate the knees of rats with osteoarthritis. Earlier stage research in mice has shown that placing tiny droplets containing the precursors to smell-sensing neurons can restore sense of smell in mice whose own olfactory neurons have been damaged. Another study took cells purified from human urine(!), turned them into iPSCs, and used them to make dental precursor cells which grew into "tooth-like structures" in mice. Fresh biological teeth for all (as opposed to the various metallic, plastic and ceramic prosthetics available currently) is surely a noble goal for dentistry and will be of particular help for elderly people struggling to chew their food.

Stem cell research is a field so vast and fast-moving that it's impossible to do it justice in a single section of one chapter of a book. By the time you read this, some details are pretty much certain to be out of date, hopefully because some of these therapies are a little closer to the clinic. It's perhaps the only aspect of aging research which gets attention and funding even barely proportional to its potential benefits. Though there are kinks to be ironed out, the pace and breadth of change is breathtaking.

You can now see why stem cells aren't the cure-all suggested by shady clinics—they aren't a generic one-shot treatment which can reverse aging on their own, but an umbrella term for a range of treatments involving different kinds of cells. Nonetheless, stem cells are soon going to be a much bigger feature of medical care, especially for the degenerative diseases of aging.

IMPROVING IMMUNITY

One place where stem cells and other rejuvenative therapies will be of use is the immune system. A good place to start is the thymus—the small organ just behind your breastbone where T cells are trained, and which undergoes a programmed decline starting in childhood. The process of useful thymic tissue turning to useless fat is "involution" and it's a surprisingly malleable process. Of the various ways to stop the thymus from involuting, or even reverse the process, probably the best-studied is sterilization: both surgical removal of the testes or ovaries and drugs that stop the action of sex hormones increase the volume of the thymus in mice.

Though it might prove difficult to get volunteers for a clinical trial of sterilization in humans, there are some fascinating studies that use historical evidence to try to work out its effects on longevity. In eighteenth-century Europe, singers called castrati dominated the opera scene. Castrati were boys castrated before puberty to preserve their unbroken voice throughout life. An analysis of their lifespan showed that it didn't differ from other male singers of the time, but the sample size was small and it's possible that some castrati hadn't actually been castrated but were just men whose voices had not broken at puberty. Another study looked at inmates of an institution for the "mentally retarded" in Kansas, at a time when the "genetically unfit" were sterilized due to policies initiated by the eugenics movement. These results were more convincing, finding that castrated inmates lived 71 years compared to 65 for other institutionalized men—and they also seemed to avoid male pattern baldness. However, there was still some doubt because 65 years was somewhat lower than general life expectancy in the U.S. at the time, which could mean the difference was an artifact of institutionalization—perhaps, for example, inmates passed over for sterilization were unhealthier, or treated differently from their castrated peers, meaning that this effect wouldn't carry over into the wider population.

The strongest evidence that castration increases lifespan in humans comes from an analysis of the eunuchs in the Korean Joseon dynasty. The Joseon ruled Korea for five centuries and the eunuchs, or **naesi,** were a key part of the Imperial court: only they, along with the **Yang-Ban,** or noble class, were allowed to become government officials, and only members of the royal family, eunuchs and women were permitted to stay inside the palace walls after sunset to preserve the royal bloodline. Around 140 eunuchs formed the **Nae-She-Bu,** an organization responsible for guarding the palace, cooking, cleaning, overseeing maintenance and running errands for the king.

The **naesi** were allowed to marry and adopt children, either girls or castrated boys, meaning, somewhat counterintuitively, that there is a eunuch family tree. In 2012, researchers used it to analyze eunuch longevity, cross-referencing with other dynastic documentation where possible to validate the data. The results were clear: the 81 eunuchs whose lifespans could be verified lived an average of 70 years, compared to three **Yang-Ban** families of similar social status whose average lifespans ranged from 51 to 56. Even the kings, who spent their whole lives inside the palace, only averaged 47 years at death—two **naesi,** In-Bo Hong and Gyeong-Heon Gi, lived under four of them, reaching 100 and 101 respectively, and Ki-Won Lee, who died at 109 years old, lived under five. This makes three centenarians out of 81 eunuchs—compared to less than one in 10,000 men making it to 100 in **modern** Japan, the country currently topping the centenarian table—and 101-year-old Gi was born in 1670, when life expectancies were decades shorter than they are today.

Unfortunately, the Joseon eunuchs' genealogical records don't make any note of thymus size. However, there's good reason to believe it's a contributing factor. The Kansas inmates lived longer primarily because of a reduced death rate from infections, suggesting involvement of the immune system. There are also experiments in mice: castrating nine-month-old mice increased the size of their thymus and improved their immune response after a flu infection, and it also dramatically improved their ability to resist cancer—when injected with tumor-inducing cells, 80 percent of control mice developed cancer, while just 30 percent of their castrated, thymus-boosted siblings did.

Evidence in mice suggests that the same holds true for females as males, but there are far less data to go on in both mice and humans due to the fact that removing the ovaries is a significantly more difficult and dangerous procedure than removing the testes. What evidence we have does point in the same direction: for example, sterilized female inmates in the Kansas study did live longer, but there were so few of them it's hard to draw firm conclusions. The effect might also be masked because the female sex hormone estrogen has a protective effect on cardiovascular health: removing the ovaries might help the thymus but lead to an increased risk of heart disease, reducing the net benefit on total lifespan.

Though sterilization is a simple intervention for lab experiments, I can't see many people queuing up to get it. Luckily, there are several alternative treatments, making use of growth hormones, stem cells or gene therapy. The hormonal approach is the most advanced, having made it as far as a small human trial conducted by a company called Intervene Immune. They gave nine men a combination of human growth hormone (HGH) together with DHEA (another hormone) and metformin (the diabetes drug and potential anti-aging pill you might remember from the last chapter) to combat the diabetes risk associated with HGH. The results were positive, and quite wide-ranging: their thymuses look less fatty on an MRI scan and they have more T cells fresh from the thymus, as you'd hope— but they also saw improvements in kidney function and, most excitingly, a reduction in their epigenetic age, as measured by the morbidly accurate epigenetic clocks we met a couple of chapters ago. This suggests that rejuvenating the thymus can go on to rejuvenate the body more generally, not just the immune system—and, given the immune system's wide remit for defense and maintenance around the body, perhaps this shouldn't come as a surprise.

There is also a more direct way to induce thymic rejuvenation, using a gene called **FOXN1.** Though it has a number of different functions in places like skin, hair and nail growth, **FOXN1** seems to be particularly critical in development of the thymus. Most babies born with DiGeorge syndrome—a genetic condition which counts among its symptoms an underdeveloped or entirely absent thymus—are missing the piece of chromosome 22 which contains the **FOXN1** gene. It's also known to reduce in activity with age in mice and humans, at the same time as the thymus disappears. Finally, and most excitingly, **FOXN1** seems to be capable of single-handedly driving thymic regeneration: researchers in Edinburgh, UK, genetically modified mice to have an extra copy of it that could be activated with a drug; giving the drug and activating the gene encouraged the thymus to regrow and produce new T cells, even in old mice. As a result, researchers are looking into gene therapy that could add an extra copy of **FOXN1** into the cells of our ailing thymuses, or drugs which could turn our existing copies back on.

The final potential approach will come as no surprise after the first part of this chapter: we could grow new thymuses using stem cells. One pioneering treatment which has been used in a handful of cases of complete DiGeorge syndrome—where a newborn's thymus is entirely absent—is a thymus transplant. The prognosis for a child with complete DiGeorge is not good: they will normally die before the age of two of infections against which, without T cells, they are unable to mount a substantial fight. A thymus transplant can improve these babies' terrible odds, and examining their blood after the operation shows that they have far more T cells. Unfortunately, the only source for thymic transplants is other babies with intact thymuses undergoing heart surgery, where the thymus is cut away to gain access to the chest, meaning there is a pretty serious shortage of supply. The obvious solution is stem cells and, while they're not ready for the clinic yet, "thymus organoids"—small, artificial thymuses grown in the lab—have been shown to work when transplanted into mice without thymuses, and rapid progress is being made in generating thymuses from iPSCs, too.

It's not obvious which of these approaches will bear fruit first, but the multitude of methods in development means that we should soon be able to stop the thymus from involuting. This will ensure that we can make fresh T cells in old age, the first step to boosting our ability to fight infections and cancer to youthful levels.

As we work toward this goal, there are other parts of the immune system which will probably need similar regeneration. One example is lymph nodes, the "glands" that sometimes swell up uncomfortably during an infection and do so increasingly infrequently in old age. The lymph nodes are where a new threat is matched to the immune cells that are best equipped to fight it, meaning that new T cells need functioning lymph nodes to properly mature, and their age-related decline inhibits our immune defenses, too. Studies show that the immune system is only as strong as its weakest link, and a revitalized thymus might not be enough to mount a strong immune response if the lymph nodes aren't in good shape. Regenerative medicine for our lymphatic system is under development, but it's at an earlier stage than work on the thymus and could do with some extra attention.

As well as looking at the training grounds of the immune system, we'll also need to look at their graduates. Cells of the adaptive immune system can be some of the oldest cells in the body—the "memory" T and B cells which stick around after an infection, ready to deploy their knowledge of a familiar foe should it return, can survive for years or even decades. This means that the cells themselves can age. The ways we'll combat that aging will probably be similar to the approaches we use throughout the body—removing senescent cells (which we've already met), and dealing with DNA damage or extending shortened telomeres (both of which we'll discuss in the next chapter).

The aspect of aging that applies to the immune system specifically is the change not in individual cells, but in the population, as a result of persistent infections like CMV. As we mentioned in Chapter 4, infection with CMV eventually leads to a bloated population of CMV-specific immune cells which clog up the immune system's memory. Up to a third of our memory T cells can be specific to CMV by old age—leaving less room for memory T cells to fight other infections (and compounded by the lack of fresh T cells emerging from the thymus).

CMV is a herpesvirus, the family of viruses behind genital herpes, cold sores, chickenpox and glandular fever. What all these diverse diseases have in common is their incredible skills when it comes to evading the immune system. While the initial infection may be obvious (like the itchy rash typical of chickenpox), the viruses go into stealth mode after symptoms subside. Your immune system never quite eradicates them all, and the stragglers can hide out in your body for the rest of your life. At times of immune weakness, perhaps a stressful life event or a severe bout of another disease, they can re-emerge. Probably the most famous herpes reincarnate is shingles—a painful, localized rash caused when chickenpox comes out of hiding. Thanks to the general decline of the immune system with age, older people are at far higher risk of shingles, as well as other latent infectious diseases.

The reason you probably hadn't heard of CMV until a couple of chapters ago is that it rarely has any symptoms—at worst, you might get some non-specific feverishness for a few days. Given its low profile, CMV is shockingly common—something like half of us have been infected by the age of 30, rising to over 70 percent by 65. (And that's in the rich world—it's pretty much 100 percent of adults in poorer countries.) It's transmitted in the body fluids of those recently infected, meaning that it's easy to catch from babies' and toddlers' saliva or, if you miss out as a kid, during sex as an adult. And most of us over 30 still have it—after infection, the virus goes to ground, biding its time. This ubiquity means that, even though you might consider it an "external" factor, it makes sense to consider CMV as a part of human aging.

Chronic CMV infection is bad news. One study found that older people with the highest level of CMV antibodies in their blood—a measure of the body's response to infection, and thus of how active CMV is in their system—were 40 percent more likely to die over the following decade than those with lower levels of antibody activity. It's not entirely clear whether this is just a correlation—perhaps CMV can flare up as a result of other underlying health problems—or whether the CMV (and the immune system's increasingly overenthusiastic response to it) is driving ill health and, ultimately, death.

How can we combat the latent threat of CMV? The first, obvious approach is that we should develop a vaccine. This would help those who haven't yet been infected and might give the rest of us an immune boost to help keep it under control. This is actually a no-brainer even if you ignore the potentially large contribution CMV makes to aging: one of the few times a CMV infection can be immediately problematic is if you catch it during pregnancy, and, worldwide, CMV is the leading cause of brain damage in children, and it can also cause other disabilities. This alone is a sufficient human and economic case for CMV vaccine research.

Another approach is to transfer some CMV-fighting cells to reinforce the ailing troops in older people, in the hope of getting the virus under control without the need for quite such an enormous population of CMV-specific T cells. This treatment has been shown to work using T cells from donors given to people undergoing an HSC transplant, and there's also been progress toward generating T cells which can target CMV and other infections from stem cells, which would be an ideal source for topping up elderly immune systems.

The final thing we could try is removing some of the CMV-obsessed T cells to free up some space in our immune memory. This leads us to the boldest suggestion I've seen to treat immune aging: a total immune reboot. This could potentially solve the problem of CMV, along with many other immune-related problems of aging. It would mean giving HSC transplants not just to people with blood cancer, but to those whose only medical problem is biological age. The reason this suggestion is bold is that HSC transplants involve wiping out the existing HSCs and immune cells, which currently means a course of chemo- or radiotherapy, something an otherwise healthy 60-something seems unlikely to sign up for. Bear with me here—it's not as mad as it sounds.

We have already discussed the classic use-case for an HSC transplant: leukemia. The blood-forming and immune cells are both killed by the cancer treatment, and an HSC transplant is then used to rebuild the blood cells and immune system from scratch. However, in recent years there's been a growing interest in using HSC transplants to treat a far wider range of diseases.

One example is multiple sclerosis (MS), where the immune system starts to destroy the myelin sheaths which protect nerve fibers, disrupting their ability to communicate. Because nerves control so many different functions around the body, symptoms can be incredibly diverse, from sight loss, to pain, to loss of motor control. MS is just one of many "auto-immune" diseases, where immune cells aberrantly learn to target the body's own cells or proteins as a threat. While there are genetic factors which can predispose you to it, it seems that MS is somewhat dependent on bad luck. For example, if one identical twin develops MS, there's "only" a 30 percent chance the other will, in spite of their identical genetics—which still isn't great odds, but it shows that there's a significant non-genetic component. Thus, if you eradicate the problematic immune cells and let the immune system have another shot at developing from scratch, you effectively give an MS sufferer another roll of the dice. For many patients, rebooting the immune system is enough to cure the disease: HSC transplants have a higher success rate than any other available treatment.

Immune reboots have been explored in severe cases of some other autoimmune conditions such as inflammatory bowel disease and lupus, with thousands of patients having undergone the procedure, and there is robust evidence that it works. There are also two recorded instances of HSC transplant seemingly curing patients of HIV, which infects the immune cells. Both had blood cancer and, when they needed a bone marrow transplant, took the opportunity to choose a donor with a mutation which makes their immune cells HIV-resistant. It worked: so far at least, neither patient has detectable levels of the virus in their blood, nor have they needed to take HIV drugs since their transplant.

In aging, everyone's immune system eventually goes wrong in multiple different ways. It's an extremely complicated process involving a rise in autoimmunity, a vicious cycle with chronic inflammation, the expansion of anti-CMV memory cells, and more. Rather than working out exactly how things got in this state, maybe it's better just to reboot the immune system and start again. In the absence of a complete understanding of how the delicate interactions between multiple populations of immune cells become imbalanced over a lifetime, could we sidestep things effectively by turning the immune system off and on again?

There is some evidence in mice that this procedure could be useful as a treatment for aging. Scientists in Texas transplanted HSCs from young mice to old ones (though without removing their old HSCs and immune cells first), and added three months to their average lifespan. Another group in Los Angeles destroyed the immune system and HSCs of old mice with radiation, then reseeded with cells from either young or old mice. The elderly mice with a new batch of young HSCs showed improved cognitive performance on a range of different tests—in many cases, comparable to young mice. Mice with

HSCs from similarly old mice showed no such improvement, continuing on a trajectory of cognitive decline like old mice which hadn't received any treatment. This suggests that a refreshed immune system may have benefits in many different places around the body.

These experiments didn't test anything specifically relating to immune function, like response to a vaccine or resistance to infection. Nonetheless, the immune system may well be behind some of the rejuvenation of these older rodents, helping with clearance of senescent cells and so on. Other contributions could come from better, healthier blood cells generally and, as we'll see in the next chapter, from beneficial signals secreted by the young stem cells. This should be enough to encourage further research into giving older humans fresh HSCs.

For now, HSC transplants are an incredibly serious procedure considered only in cases where there are few other options. This is because the "conditioning" chemo- or radiotherapy that comes first can be punishing, and spending weeks without an immune system while waiting for it to rebuild from scratch puts patients at serious risk of infection (not least from a re-emergence of latent viruses like CMV). However, I think it also provides a fascinating thought experiment when transposed into the context of aging. As clinical understanding of HSC transplants has improved, mortality in MS patients undergoing this procedure has dropped to

0.3 percent. That's not **that** low—doctors and patients would be understandably hesitant about undergoing a procedure with serious side effects and a 1-in-300 chance of death for anything other than a pretty substantial benefit. However, putting this risk in the context of aging could change your outlook: in the rich world, your annual chance of death exceeds 0.3 percent around the age of 50. So, beyond that age, might a 0.3 percent chance of death from a procedure be worth it if it reduced your chance of death overall by more than that? We don't have the evidence for giving HSC transplants for otherwise healthy 50-somethings, of course—this is just an example to show how thinking about things from the perspective of aging could redefine what we mean by a "risky" procedure. And, in the long run, we won't need to bludgeon the immune system to death with broad-brush, side-effect-laden treatments like chemotherapy anyway: given its benefits to people with autoimmune diseases, work is ongoing to make HSC transplants safer.

It's not hard to imagine that a rejuvenated thymus, improved lymph node performance and a fresh batch of HSCs could significantly improve immune function in the elderly. There are results in mice, companies and ongoing clinical trials, all of which point in the right direction. The upshot is likely to be worth it, not just in terms of fighting infections, but also to reinforce the immune system's many other roles around the body, including removing senescent cells and catching early-stage cancer before it becomes a problem.

MODIFYING THE MICROBIOME

Given the synergistic relationship between our immune cells and gut flora, another thing an improved immune system might help with is our microbiome, maintaining its balance later into life and reducing chronic inflammation. However, as we saw when introducing the microbiome as the new kid among the hallmarks of aging, we are only just beginning to understand the many factors that affect the health of our microbes, and how their health affects ours. There are probably situations where the quickest intervention is one which deals with the microbiome directly, allowing beneficial bacteria to help the rest of our body out.

There are various ways to try to restore balance to the microbiome. The simplest is probiotics, which you might be familiar with from the dairy aisle of your local supermarket. Probiotics seek to introduce live microorganisms into your guts when you simply eat or drink a product containing them. Another potential avenue is consuming prebiotics— substances which are undigestible to us, but delicious to good bacteria in our guts. The current candidates are mainly various chains of sugars, oligosaccharides and polysaccharides. Like probiotics, they have been shown to alter gut microbial populations in a good way. Pre- and probiotics can be combined, as "synbiotics," effectively giving good bacteria together with a starter pack of nutrients to get them going.

More work is needed to establish which pro-, pre- and synbiotics will be most effective in different scenarios, but progress is rapid. Small studies have demonstrated that various drinks, biscuits and capsules containing cultures of bifidobacteria and lactobacilli are able to improve numbers of those beneficial bacteria in the intestines of elderly volunteers and push out problematic ones—and suggest that these treatments can have beneficial effects on the immune system, too. A probiotic cocktail of nine different strains of bacteria known as SLAB51 was able to damp inflammation, reduce beta amyloid and tau aggregation, reduce levels of advanced glycation end products and slow cognitive decline in mouse models of Alzheimer's. Probiotics, prebiotics and synbiotics have also been successfully used in small human trials to improve symptoms in Alzheimer's and control sugar in pre-diabetes.

As our understanding of the microbiome advances, and more is known about how these helpful and harmful bacteria and fungi affect our bodies in fitness, fatness, health and aging, we will be better able to identify which bacteria and other microorganisms will do the most good when transplanted. But the most exciting current results come from a more ambitious technique: a full microbiome transplant.

The story starts with the turquoise killifish, one of the world's shortest-lived vertebrates (vertebrates are animals with a backbone, like us) because it ekes out an existence in "ephemeral pools" in Zimbabwe and Mozambique. These seasonal ponds—often little more than puddles—see birth, mating and then death of little populations of killifish, whose eggs must then survive long months baking in the dry mud and waiting for rain to fill the pools and start the cycle of life again. This makes it a good potential model organism for aging research—not as wildly different from us as a fly or a worm (and, critically in this case, with a rich ecosystem of gut flora far more like ours than a fly's), but with a convenient few-month lifespan which means experiments can be completed on a user-friendly timescale.

Researchers used the killifish to investigate the effects of a microbiome transfer between young and old fish. Two-month-old fish had their own middle-aged microbiome eradicated with a cocktail of four potent antibiotics, and were then given youthful gut microbes to replace them. Not only did they live longer—their lifespan was more than five months on average, a 25 percent increase over fellow fish with regular gut flora—but they also darted around the tank more frequently in old age, in a kind of fishy proxy for delayed frailty.

There is also some early work in mice suggesting that younger gut flora can improve health in aging mice. An experiment on mice with accelerated aging found that a microbiome transplant from normal mice was able to extend their lifespan by around 10 percent, and that supplementing them with a single species of bacteria which declines with age in both mice and humans extended their lives by 5 percent. In Chapter 4 we discussed how putting old and young mice into cages together and allowing them to eat one another's feces—thus performing a de facto microbiome transplant—worsened inflammation in young mice given old microbes. The converse was also true, with old mice sharing a cage with younger ones seeing a boost to their immune systems. Slightly confusingly, follow-up experiments showed that actively transferring the microbiome between mice (rather than just letting them eat one another's feces) improved immunity whether from young to old or old to young. Clearly, transferring gut flora can have a substantial impact on health—but more work is needed to establish exactly which bugs are beneficial under what circumstances.

Humans can have microbiome transplants, too—they involve extracting fecal material (so, poo), purifying it and then either inserting it via colonoscopy or enema or asking the patient to swallow a capsule filled with freeze-dried powder. If the idea of an enema of someone else's feces fills you with horror, you should know that they're already used as a treatment for **Clostridioides difficile** infections (also known, more pronounceably, as **C. diff**), where the assistance of gut flora from a healthy donor can help to eradicate the intruding bacteria. If the alternative is continuing to struggle with a serious infection, a poo transplant is probably preferable, and it might be worth banishing our squeamishness if the alternative is worsened aging, too. Microbiome transplants have also been shown to help with obesity, diabetes and Parkinson's in mice, and human trials are following up on these results.

Since one of the benefits of our microbiome is providing us with beneficial molecules, the final approach to realizing its benefits is to identify these bacterial by-products, cut out the middlemicrobes and deliver them directly as medication. A recent investigation of this idea was an audaciously exhaustive experiment in our favorite nematode worms, **C. elegans.** The worms are usually grown on plates of **E. coli** bacteria, which serve both as their food and the only organism in their gut microflora, making their intestines' inhabitants substantially easier to study than the rich ecosystems found in humans or fish. **E. coli** is a very common lab organism and, as such, it's possible to order strains off the shelf which are missing any of the 3,983 genes whose absence doesn't kill the bacteria. Thus, the worm study involved growing 3,983 plates of worms, each eating **E. coli** which were missing a different gene. Of these mutant bacteria, 29 increased the lifespan of the worms. Of those 29, 2 of the longevity-promoting genes were found to control production of a polysaccharide called colanic acid. Finally, giving the worms colanic acid directly increased their lifespan by 10 percent, even when living on normal **E. coli.**

As well as providing proof of the principle that directly consuming microbial metabolites can increase lifespan, this study is also a showcase for the incredible usefulness of nematodes for massive, systematic searches like this—performing thousands of parallel experiments with different bacterial strains in mice simply wouldn't be possible. The simplest outcome would be if colanic acid has some direct application in higher organisms like us—but, regardless, if this general principle holds up, it will be a race to see whether molecules or microbes are the easiest way to realize the benefits of our gut bacteria in the clinic.

While it's too early to be sure whether microbiome-modifying medicines will find a place in fighting aging, it certainly seems plausible. With a better understanding of how these helpful and harmful organisms affect our bodies, we might all be periodically popping tablets of freeze-dried fecal matter to keep our guts in top condition.

KEEPING PROTEIN PRISTINE

Collagen is a structural protein whose name you may recognize from (often questionable) claims on the labels of skin and face creams. As with many not-quite-scientific claims, there is a grain of truth here: collagen is the most important protein in the structure of skin, and indeed many other tissues in the body, from blood vessels to bones. It's our most abundant protein, contributing two or three kilos to the average adult's bodyweight, and lasts for a surprisingly long time: current estimates suggest that it takes years for the collagen in your skin to be "turned over" (the biological term for being broken down and replaced), and collagen in cartilage, which provides the smooth padding between bones at our joints, may last a lifetime.

A single molecule of collagen looks like a really tiny piece of string, made up of three strands of atoms twisted around one another. The collagen molecules are then held together by crosslinks which clip on to specific points on a collagen molecule and attach it to its neighbors. This is a fibril which, if the individual collagen molecules are string, is a thick, long section of rope. These fibrils then bind together, along with various other molecules, to form even thicker structures called fibers—like the thick, many-stranded cables which hold up a suspension bridge. The exact structure of individual collagen molecules is absolutely critical to the multi-thousand-molecule megastructure of a collagen fiber. It dictates how the molecules coalesce into fibrils, and how the fibrils assemble into fibers, and what other molecules are drawn in to act as support, glue or lubricant. The result, in turn, controls a fiber's properties—not too stiff, not too flexible, but just right in a huge range of biological contexts. The same basic molecular building blocks can make different types of collagen ranging from stretchy in skin and blood vessels to tougher in tendons and strong and load-bearing in bones. This is an oft-neglected biological miracle: we rarely step back to admire the exquisite evolutionary engineering which allows proteins to self-assemble into massive, incredibly effective teams.

Unfortunately, this intricate structure can be disturbed by chemical modifications which change the structure of the individual collagen molecules that make it up. Highly reactive chemicals like sugars and oxygen can stick to the collagen, causing widespread disruption. Dangling sugars can crowbar open fibrils, allowing water to rush in and unbalance their carefully calibrated interior chemistry. Many of these sugary modifications are somewhat transient, and can just fall back off, but occasionally these modifications can themselves be modified. This can eventually result in an advanced glycation end product, or AGE, which is permanent. AGEs can dangle from a single collagen molecule, just like their sugary precursors, or they can crosslink collagen molecules, handcuffing two proteins together and stopping them from moving smoothly past one another.

All of these changes can also disrupt the purpose-built crosslinks whose type and frequency dictate the collagen's mechanical properties. The large-scale effect of these microscopic modifications is to shift collagen away from the sweet spot between stiffness and stretchiness. Though the effect varies by tissue, the most common is a reduction in elasticity, which you can easily see for yourself: a pinched patch of skin springs back into position ever more slowly with the advancing years.

As well as their direct effects on the collagen itself, these chemical changes can set up feedback loops that make matters worse. Collagen provides the scaffolding for many cells to sit on, from skin to bone. In turn, the cells are responsible for the collagen's upkeep, producing new collagen to renew the scaffolding on which they sit, like responsible citizens keeping their neighborhood in good repair. The cells bind to collagen in precisely defined sites, whose positions and properties are another feature specified by collagen's molecular structure. As this structure is disrupted, these binding sites can be obscured or reduced in stickiness, meaning that cells are attached less firmly in place. This is bad in itself, reducing the integrity of the tissue, but perhaps worse is how the cell reacts. As its attachment to collagen becomes weaker, it starts to ponder its identity. When cells are deciding how to behave, some of the cues which guide them come from the "extracellular matrix," or ECM, it's stuck to. Where firm contact with collagen reaffirms a cell's purpose as a skin cell or a cell in an artery wall, the loss of that contact introduces some doubt. Counterintuitively, this means that it produces **less** new collagen: rather than worrying about the lack of ECM and producing more to compensate as you might expect, the cell becomes convinced by its absence that it's not one of those cells which sits on a collagen scaffold, and so it has no need to produce any.

It's also known that cells can detect AGEs, using receptors on their surface which are known by the acronym RAGE—for "receptor for advanced glycation end products." (Protein modification definitely has the best acronyms in biogerontology.) Activation of these receptors promotes inflammation and cellular senescence, and it's not entirely clear why; one idea is that the cells are calling the immune system for help to clear up AGE-damaged collagen, but there isn't much evidence to guide us one way or the other. Nonetheless, this means that these damaged proteins can cause cells to contribute to chronic inflammation which, as we've seen, is behind many of the processes in aging.

All of this means that, as collagen undergoes chemical damage throughout our lives, the structural integrity of the ECM begins to fail. Exactly how much blame to assign to the different contributors— glycation, oxidation, AGEs and cells' responses to all of the above—remains unclear, but the consequences are well understood: stiffer and weaker skin, arteries, lungs, tendons and so on, causing a reduction in tissue integrity all around the body.

What to do about these modified proteins presents something of a challenge. The prevailing theory has long been that the major problem is AGEs, and specifically the crosslinks they form between collagen molecules shackling them together and reducing flexibility. However, work from just the last few years is calling this received wisdom into question— it seems that AGEs, sadly for their catchy acronym, may only be part of the story. Other modifications which mostly don't crosslink the collagen might be more important than AGEs. Other types of glycation are more abundant, and their power to disrupt natural crosslinks probably has a larger effect than a few extra AGE-related ones. The overall picture is one of chemistry falling out of balance: collagen getting sugarier and more oxidized with age, and losing its characteristic properties as a result—a rather more subtle collection of multiple phenomena, not just stiffening caused by AGE crosslinks. This has taken a long time to uncover both because the experiments are technically difficult and because the work is interdisciplinary, requiring careful collaboration between chemists and biologists, and such work is often overlooked by funders.

Given the prominence of AGE crosslinks in previous research, most ideas for reversing collagen aging have focused on getting rid of them. Scientists are working on "AGE-breaker" drugs that can cleave the modifications from the collagen in the hope of restoring its youthful suppleness. Though our new, nuanced understanding of collagen aging calls the logic of AGE-breakers into doubt, it's probably still worth trying them. As we've mentioned before, the surest way to test if something is an important contributor to aging is to get rid of it and see what happens. If we can reliably remove AGE crosslinks and it works, great; if tissue flexibility doesn't change, we might need to enact plan B. This could build on work done developing the AGE-breakers, modifying whatever is found to be effective against AGEs to instead chop off other modifications that are more biologically relevant.

However, the most promising approach might be to avoid meddling with this complex chemistry altogether. We could sidestep our imperfect understanding entirely if we could encourage our bodies to rip up old collagen and replace it from scratch. This is clearly possible in principle—our bodies must have built it in the first place. They are also quite capable of regenerating collagen in many places, albeit not at the required rate: though some collagen lasts a lifetime, careful measurements of collagen longevity in mice suggest that there are some places where it lasts only weeks rather than decades—there's no hard-and-fast biological reason that we couldn't emulate our furry counterparts. The good news is that we can stimulate collagen turnover to some extent with exercise: exercise causes mild collagen damage, setting the body's natural repair and replacement processes in motion. Unfortunately, exercise can only take us so far, and we're not sure how we might take a more comprehensive medical approach to encouraging cells to ramp up the destruction and rebuilding of the matrix around them.

Optimistically, there's some reason to hope that tackling other hallmarks of aging could have a positive effect on our collagen. First, many of the sugar-related reactions with proteins are intrinsically reversible. The increased blood sugar levels associated with both aging and diabetes mean that sugars are more likely to stick than to fall off, which causes an increase in the number of glycated proteins; better control of blood sugar could turn this process around, allowing the collagen to recover by simple chemistry. Next, the SASP secreted by senescent cells is partly made up of enzymes which degrade the extracellular matrix, and there's also evidence that some dysfunctional immune cells called neutrophils rampage through it in old age, leaving a trail of destruction in their wake. Thus, removing senescent cells and rejuvenating the immune system may at least slow the damage to the proteins outside our cells. If we get luckier, restoring our bodies to an otherwise youthful state may cause cells which previously did perform adequate levels of collagen maintenance to get back to work—though there may still be some places, like cartilage, where turnover is too slow even at youthful rates.

This is one of the aspects of aging whose solution seems the least certain at the moment. More research is needed to work out exactly what happens to long-lived proteins outside cells and how to fix them. There are also other proteins that this affects that we've not taken much time to discuss: skin and arteries both have another major structural constit-uent called "elastin" which, as the name suggests, is partly responsible for tissue flexibility; aging of pro-teins in the lens of the eye leads to loss of both flexi-bility and transparency in old age; and more besides.

If we had to choose one area on which to focus this research, I would make the case for the collagen and elastin in the blood vessels—high blood pressure is a leading cause of death, disease and dementia, and degradation of the ECM there almost certainly has the biggest impact on our health. Though improv-ing the collagen in our skin could restore some of its youthful sheen, I would far rather have saggy skin and young arteries than the reverse. We could then turn the tools and techniques developed to revital-ize our circulatory system to the other places in our bodies where proteins are modified.

7

Running Repairs

Sometimes the best approach to fixing a hallmark of aging is neither removal nor replacement, but repair. Our DNA is a case in point: our cells wouldn't last long without their molecular instructions, and replacing the two meters of the stuff in tens of trillions of cells would be somewhere between impractical and impossible. That means, from trimmed telomeres to mutations, we're going to have to find ways to fix DNA while it's still in our cells.

We'll also take a look at rebalancing the signals in our blood to restore them to more youthful levels and patching up damaged mitochondria such that they can keep generating energy for our cells into old age. We'll start with telomeres.

TELOMERE EXTENSIONS

Every time a cell divides, its telomeres get shorter. Since many of our tissues rely on dividing cells to replenish them, their telomeres eventually become "critically short," leading them to cell suicide or senescence. People with shorter telomeres tend to die sooner than those with longer ones. Is there a way to undo the erosion of our DNA's protective end caps and, in doing so, extend our lives?

The telomere story starts in 1984, when scientists Elizabeth Blackburn and Carol Greider were investigating them in a single-celled pond creature called **Tetrahymena.** These are tiny organisms, covered in so many microscopic hair-like protrusions that they look furry. My favorite fact about them is that there are seven different **Tetrahymena** sexes, which the critters decide at random during mating—resulting in 21 different permutations of parentage, and children which can then be any one of those seven sexes again. Blackburn noticed that, under certain conditions, **Terahymena**'s telomeres could grow. This seemed strange: the expectation at the time was that DNA was a constant, unchanging blueprint for an organism, not something which could be added to at will. How and why were these tiny creatures making more of it?

Tetrahymena's most useful quirk for telomere research is that they have around 20,000 chromosomes per cell, meaning 40,000 telomeres to probe—rather more than the paltry 46 chromosomes and 92 telomeres in a human cell. Thus, Blackburn reasoned, if there was some mechanism which extended telomeres, these hairy-lookin' microbes should be buzzing with it. After years of painstaking research, Greider and Blackburn finally isolated the enzyme responsible for topping up **Tetrahymena**'s telomeres. They christened it telomerase, and it's turned out to be rather a big deal: the pair were awarded the Nobel Prize in Physiology or Medicine in 2009, along with Jack Szostak, who helped Blackburn demonstrate the protective effects of telomeres with experiments in yeast.

Telomerase seemed to be the immortality enzyme, for cells at least. Disabling the gene in **Tetrahymena** made the little seven-sexed cells, which would normally reproduce indefinitely, die within a week. Most animal cells don't have active telomerase, and can be used for the converse experiment: adding an extra copy of a telomerase gene allows them to divide indefinitely, sidestepping senescence. This was first done in human cells in the mid-1990s, at a biotech company called Geron—ironically, using cells belonging to Leonard Hayflick. Hayflick made a serendipitous donation of skin cells from his leg: he was showing a TV crew filming a documentary about his work how to take a sample of skin from which to culture cells, and happened to ask Geron's chief scientific officer, Mike West, for a scalpel. West thought the opportunity to measure the "true Hayflick limit"—how many times Hayflick's own cells would divide before turning senescent—was too good to pass up. With fortunate timing, scientists at Geron had recently isolated the human gene for telomerase, so West decided an even better experiment would be to insert an extra copy of the gene into Hayflick's skin cells and see what happened. Hayflick's unmodified cells hit their eponymous limit on cue, but those with extra telomerase just kept dividing, making them the first human cells to be "immortalized" with telomerase. The irony is magnified because Hayflick—now in his nineties—is a lifelong skeptic that we will ever be able to intervene in the aging process.

This miraculous behavior raises an obvious question: could telomerase do for whole humans what it does for cells in a dish? Judging by articles in the popular press in the 1990s, you'd be forgiven for thinking so. This tale of telomeres as a simple cell division clock, and telomerase as a way to restore them, is so easy to understand that it multiplied like cells with active telomerase. The fact that we're not all already on telomere-enhancing pills probably tells you that it turned out to be a little more complicated than that.

The most obvious problem is cancer. To form a tumor, a cancer cell needs to divide over and over again, which means it needs to stop its telomeres from getting critically short. As a result, almost 90 percent of cancers reactivate telomerase in order to sidestep cellular senescence. (The other 10 percent use a mechanism known as ALT, which stands for "alternative lengthening of telomeres"—an acronym which, like "dark matter" and "dark energy" in astrophysics, exists primarily to cover for the fact that we have very little idea what it is or how it works.) Active telomerase isn't enough in itself to turn a cell cancerous, but we'd rather not pre-tick any boxes on cancer's checklist if we can avoid it.

This worry was borne out by the first experiments using telomerase in organisms more complex than **Tetrahymena.** Scientists added extra copies of the telomerase gene into mice and, while some benefits were observed—including thicker skin and faster-growing hair—so, too, was an increased risk of cancer. Studies doing the opposite, removing the natural telomerase gene from mice, found that a lack of the enzyme suppressed tumor growth. Thus, it seemed pretty clear that telomerase was a pro-cancer enzyme, and this realization burst the telomerase bubble somewhat.

Thus, telomeres seem to be a key component of our cellular anti-cancer mechanisms. As well as stopping the ends of chromosomes from being glued together, and protecting the important parts of our DNA from being chopped during cell division, as we learned about in Chapter 4, they have also been put to use protecting whole organisms from cancer by the enterprising hand of evolution. By keeping count of how many times a cell has divided, telomeres provide a mechanism to catch cells which have done so too many times. If a cell runs out of telomere and ends up replicatively senescent, it could save your life.

This is why telomerase is disabled in most adult human cells.* However, the enzyme clearly can't be gotten rid of entirely, even in non-cancer contexts. For example, it's imperative that embryos have the ability to regrow telomeres between generations, lest children's lives be stunted by their parents' short telomeres and the species go extinct. Pluripotent stem cells, whether embryonic or induced, continually use it to keep their telomeres long, allowing them to divide indefinitely. It's also active in some adult stem cells, like blood-manufacturing HSCs, but only enough to slow rather than entirely forestall telomere shortening, and sometimes in T cells at times of infection, when they need to rapidly multiply in numbers to face down the specific foe they are tailored to target.

Evolution seems to be playing a deft hand optimizing use of telomerase. The dynamics of telomeres, as so often, are a biological trade-off between avoiding aging and avoiding cancer.

The most extreme example is a rare genetic

---

* Though telomerase is disabled in adult **human** cells, things get a bit more complicated in other species, ranging from short-lived mice to long-lived, cancer-free naked mole-rats, which have active telomerase in their cells. Different species have found quite different ways to balance on the telomerase tightrope.

disease named dyskeratosis congenita, or DC, which we now know is caused by sufferers having very short telomeres. Patients have problems with fast-dividing tissues, like skin, hair and blood, and experience something a bit like accelerated aging, with rapid hair graying, lung problems and osteoporosis. There's even a dark irony, in that DC patients are more susceptible to certain kinds of cancer: this is because really short telomeres induce a state called "crisis" and, if a cell fails to go senescent, the chaos in its DNA can induce cancer-causing mutations; and a lack of telomerase weakens our immune system which might otherwise catch it early.

At the opposite extreme, a family was found in Germany with a mutation affecting a single DNA letter 57 bases before the start of the telomerase gene. This increased the amount of telomerase some of their cells produced by around 50 percent and gave them a dramatically increased risk of cancer. Four of the five family members carrying this mutation had developed melanoma and the other, at 36 years old, was reported to have several worrying moles on their skin. Another developed melanoma at 20, followed by cancer in her ovaries, then kidneys, then bladder, then breast, and finally lung cancer, which killed her at the age of 50. It really is incredible the amount of trouble a single DNA base in three billion can cause.

Thus, telomerase is a Goldilocks enzyme—too little, and your fast-dividing tissues fall apart;

too much, and cancer finds it all too easy to take hold. Thankfully, most of us have it pretty close to just right. While there is a bit of natural variation in the population which means that we each have slightly different levels of telomerase, it doesn't much matter overall. If you look across human populations, you can compare people with subtle DNA variations which slightly increase or decrease telomerase activity: more active telomerase slightly increases the risk of dying from cancer, but doesn't make much difference to the risk of dying overall because you're slightly protected from other problems like heart disease which are associated with short telomeres.

So, if telomere length and telomerase levels are rather like walking a tightrope between aging and cancer, what practical interventions are there which might help us stay on the tightrope rather than falling off in either direction? If you want to follow the story of efforts to turn telomerase into a therapy, you could do worse than follow the career of molecular biologist María Blasco. In 1993, she moved from the lab in Spain where she'd completed her PhD to the U.S., to work as a post-doc for Carol Greider (whom you'll remember from her exploits discovering telomerase in **Tetrahymena**).

Blasco wasn't deterred when the telomerase bubble seemed to burst in the early 2000s. Convinced that understanding telomeres could lead to new medicines for diseases caused when they get too short, her lab continued experiments on the enzyme. In 2008, her group published a paper showing that telomerase could extend lifespan in mice—as long as they were also genetically modified to be cancer-resistant. Mice engineered to have both extra telomerase and three additional DNA-defending genes, which encourage cells to die or go senescent if they've got pre-cancerous mutations, lived 40 percent longer on average than unmodified mice. This offers a ray of hope—the battle between cancer and aging seems not to be a zero-sum game, where every victory over one results in death by the other. In a complex biological system, ramping up two competing effects can—and in this case, does—synergize to give a net benefit.

A follow-up experiment tried a different kind of gene therapy in adult mice. The mice were injected with billions of viruses* which, rather than causing an infection, delivered an extra, temporary telomerase gene to their cells. Those injected aged one (roughly equivalent to 40 in human years) lived an average of 20 percent longer than their peers. The telomerase-treated mice also saw an uptick in their health: better control of blood sugar, higher bone

---

* Adeno-associated virus, or AAV, if you're interested—a common "viral vector" used in genetic modification in the lab, and a leading candidate for delivering human gene therapy.

density, plumper skin and improved performance walking a (literal) tightrope. Most important of all, there was no obvious increase in risk of cancer.

This is promising and could feasibly be deployed in adult humans—but, as the saying goes, mice aren't just tiny people. One potential difficulty translating this finding in mice into humans is that mice are considerably shorter-lived—where the average telomerase-treated mouse in this study lived another year and a half post-treatment, a human treated at the equivalent age would have decades of remaining lifespan to accrue cancer-causing mutations. Could this mean telomerase is safe for short-lived mice, but hazardous for long-lived people?

To head off this criticism, Blasco's lab tried the same viral gene therapy in adult mice genetically modified to have massively increased cancer susceptibility. There was no discernible difference in the cancer rates between mice given extra telomerase and those given a control dose of virus which didn't contain any DNA—both were equally, and appallingly, high. This suggests that, even in a highly cancer-prone environment, this kind of telomerase gene therapy at least doesn't worsen matters—and is suggestive that gene therapy in adult humans might not be as carcinogenic as was originally feared.

One final experiment from Blasco's lab involved making mice which had very long telomeres, but completely normal telomerase. These mice lived 13 percent longer on average than mice with normal-length telomeres. They also saw a number of health benefits: lower bodyweight, lower cholesterol, less DNA damage and, crucially, less risk of cancer. This experiment suggests that very long telomeres aren't intrinsically problematic, but it's hyperactive telomer**ase** that increases cancer risk (as we saw in the German family). Thus, if we could boost telomere length without turning up our existing telomerase genes, perhaps we could side-step the telomerase trade-off between cancer and degeneration altogether.

With all this fascinating evidence in mice, what's needed now is more trials to check out telomerase in animals more like us. One option might be to jump straight to humans. As we've seen with other hallmarks, there are plenty of cases where telomeres are responsible for diseases more acute than slowly aging to death. One place to start could be people suffering from dyskeratosis congenita, or a handful of related diseases, whose direct cause is insufficient telomerase.

Another possibility is idiopathic pulmonary fibrosis, the lung disease which you might remember as the first place we tried out senolytic therapy. (It's reassuring that short telomeres and senescent cells turn up in the same places, given that one can cause the other.) Experiments with telomerase gene therapy in mice suggest that it can reverse IPF and, given that patients don't currently have any good treatment options, some would probably be willing to take a punt on telomerase. Patients enrolled in these trials will be watched very carefully for any increase in cancer risk and, if this doesn't materialize, we could start to prescribe them to broader groups of people. Since shorter telomeres put people at risk of cardiovascular disease, heart patients could come next. If people at risk of heart trouble given telomere therapy don't start to die of cancer, we can imagine telomerase being prescribed prophylactically to all of us.

Gene therapy isn't the only option. We could also look for drugs or supplements which naturally enhance the activation of the telomerase genes already present in our cells, as long as their action is transitory. The most studied is TA-65, a chemical isolated from a herb used in traditional Chinese medicine, which can extend both telomeres and healthspan (but not lifespan) in mice by activating telomerase but without increasing their cancer risk, and there's some evidence it might have positive effects on health in people, too. It would certainly be worth rifling through drugmakers' back catalogues to see if there are any other molecules which could be put to use.

So the immortality enzyme of the nineties turned unloved causer of cancer in the oughts is staging a comeback. Experiments in mice present an increasingly watertight case that telomerase, wielded judiciously, need not be a double-edged sword, and there's no obvious impediment to trying these therapies out in people. And, if they work, we can move from walking the telomerase tightrope with cancer on one side and degenerative diseases on the other, to dancing along it, protected from both.

CAN YOUNG BLOOD TEACH OLD CELLS NEW TRICKS?

In aging biology, the prize for the experiment with the richest overtones of Gothic horror must go to heterochronic parabiosis. The technique somehow manages to combine Frankensteinian sewing together of body parts with a vampiric taste for young blood. In such studies, two animals (usually rats or mice) of differing ages have the skin on one side of their body peeled back, and their two exposed flanks are then sewn together. As the healing process gets under way, tiny blood vessels grow between their bodies until the artificially conjoined twins share a blood supply.

While it may sound gruesome, heterochronic parabiosis has given scientists a new way to understand, and potentially devise treatments for, the aging process. One of the many vital functions of blood is to act as a telecommunications network for the body, ferrying dissolved chemicals which act as messengers that can affect how cells behave throughout the body. Watching what happens when a young mouse is confronted with old blood, or an old mouse is given a refreshing blast of the young stuff, has given us fresh insight into how the effect of systemic, internal factors can drive aging. It also provides inspiration for new treatments—which thankfully won't involve sewing old people to teenagers.

Parabiosis was first developed purely as a scientific novelty in the nineteenth century. In 1864, physiologist Paul Bert sewed two rats together and demonstrated that they shared a circulatory system by injecting one with deadly nightshade. Deadly nightshade is also known as belladonna—from the Italian **bella donna,** meaning "beautiful woman"— because, during the Renaissance, women used eye drops made from its berries to dilate their pupils and make them appear more attractive (its English name should tell you why this was a terrible idea). After injecting it into one rat, its pupils dilated rapidly; within five minutes, the other rat's pupils enlarged, too, showing that it had made its way into the other rat's blood, and proving that their circulation was shared.

Parabiosis has since been used to study obesity, cancer and even tooth decay—scientists can change some factors for one animal in the pair, while parabiosis makes sure that their internal environment is mostly shared, allowing the effects of the changes to be teased apart. The tooth decay experiment is a neat example. Researchers in the 1950s wanted to work out whether it was sugar's direct effects in the mouth or its indirect effects in the blood that caused rotten teeth. They turned to parabiosis, feeding one rat in the parabiotic pair a sugary diet and the other normal food. The shared blood supply meant that both rats had equally sugary blood—but only the one actually eating the sugar developed cavities, proving that blood sugar levels aren't a factor. It may be a bit grisly, but it's an elegant way to ensure a fair test.

Parabiosis means, perhaps a little euphemistically, "living beside." For aging researchers, the interest is in the "heterochronic" version—when you sew together animals of different ages. The first experiments of this type also took place in the fifties, performed by Clive McCay (the dietary restriction pioneer you will remember from Chapter 3). He and his team conjoined 69 pairs of rats in total, in what now seem like rather primitive procedures with varying degrees of success. Eleven pairs died within weeks of "parabiotic disease," thought to be a consequence of both bodies' immune systems going to war against the foreign tissue in the other (interestingly, we still don't know exactly what causes this—but it's far less common in modern experiments, probably due to improvements in sterile surgical techniques). Other pairings met their end when one rat chewed the head off its partner (modern incarnations allow the animals a couple of weeks in the same cage to get used to each other before attaching them, for both practical and ethical reasons). The results were suggestive, with the older animals in heterochronic pairs showing improved bone density, but the experiments weren't systematic enough to be truly convincing.

Experiments in the early seventies provided a more robust picture. Scientists compared the lifespan of pairs of rats joined heterochronically to both isochronic (same-aged) pairs and animals living a more conventional solo existence. The lone animals lived around two years; rats in isochronic parabiosis lived slightly less long, confirming (perhaps unsurprisingly) that being sewn to another rat is a physically stressful procedure; but the older rat in a heterochronic pairing lived longer—about the same as a solo rat if the pair were male (meaning that being attached to a younger partner was enough to cancel out the disadvantages of the parabiosis itself), and three months **longer** than normal if the pair were female.

What's shocking is that, after these early results, parabiosis went the way of so much promising early aging research and the field lay more or less fallow for the next 30 years. It wasn't until the early 2000s that it was finally rebooted, by wife-and-husband team Irina and Michael Conboy. The 1970s research had left key questions unanswered: it's all well and good to show that rats live longer when sewn to a younger partner, but what drives this increased longevity? The Conboys were interested in one aspect in particular: how the decline in stem cell function with age affects the ability of tissues to regenerate. To what extent is this decline driven by the aged environment of an old mouse body as opposed to any intrinsic problems in the cells themselves?

As we get older, it takes us longer to recover from injuries, be they cuts and scrapes or broken bones. As we've already discussed, much of this is because the stem cells which would normally replenish these tissues slowly decline in function—fewer or less enthusiastic stem cells produce fewer progenitor cells which can go on to replace those damaged or lost in an injury. The same is true of old mice, so the Conboys chose to see what happened to mouse rates of healing in various parabiotic combinations— young-to-young, old-to-old, and old-to-young.

Looking at three different tissues—muscle, liver and brain—their results were clear-cut. Old mice attached to young mice healed as well as a young mouse attached to another young one. As proof that it was something to do with signals in the blood reactivating the older mouse's cells, rather than youthful stem cells helpfully carried by the blood on a rescue mission from the younger partner, they genetically engineered some of the young mice in the experiment so that their cells glowed green. On investigating the healing tissues under the microscope, only 0.1 percent bore the distinctive green glow—pretty much the whole healing effect came from reawakening dormant cells in the older mouse.* Confirmatory experiments took cell samples from old mice in a dish and bathed them in youthful blood plasma—the straw-colored liquid part of blood, sieved of its cells. The results were much the same: young plasma rejuvenated the old cells, restoring their potential for growth.

These results are genuinely remarkable. Old cells aren't irreversibly decrepit, damaged beyond all hope of repair—instead, there's latent capacity, able to be eked out by the rejuvenative power of a youthful partner. The ability of old cells and organs to perk up when provided with an improved environment wasn't a given—it wouldn't have been surprising to find that they were intrinsically worn out, unable to flourish even with encouragement. Instead, an old mouse can be rejuvenated by attaching it to a younger one, living longer and healthier by having **its own cells** reawakened by a more youthful signaling environment.

The message the press took away from this was even more compelling: young blood has regenerative powers. Not only could this be a miracle cure but, as an added bonus, it taps into centuries of vampire folklore: suddenly, drinking the blood of virgins doesn't seem quite so far-fetched as a strategy

---

* The gene responsible for green fluorescent protein, or GFP, was first isolated from a jellyfish in the 1990s. Since then it—and modified versions which glow in other colors, with delightful names like mCherry, T-Sapphire and Neptune—have become indispensable

for immortality. The study, published in 2005, made headlines around the world.

Unfortunately, this being biology, things aren't quite that simple. First, for anyone thinking of drinking young blood, you should be aware that enzymes in your stomach will thoroughly break down most of the signaling molecules it carries before they ever make it into your own circulation. This means sucking blood from someone's jugular won't be of any use. However, it's not just the traditional vampiric mode of delivery which is flawed. What also didn't make so many headlines was the significant negative effect on the younger mice in the partnership. This suggests an alternative explanation—rather than young blood being the elixir of life, maybe older blood is deadly, and the service the young mouse is providing is dilution of the problematic signals in older blood, at great cost to its own health. (In fact, it's probably a bit of both.)

The final caveat is that heterochronic parabiosis is far more than just mixing blood. The older animal has the privilege of the younger one's youthful organs. The younger rat or mouse has a better liver and kidneys for filtering toxins; better lungs and a stronger heart to make sure more oxygen is delivered to the organs of both mice; a youthful immune

---

tools in biology. Their distinctive glow under a microscope renders what could otherwise be very complex experiments—like determining which mouse two basically identical-looking cells came from—incredibly simple.

system with a fully functional thymus, better at seeking out and destroying bacteria, viruses and pre-cancerous or senescent cells; and so on. There are also far more everyday factors: for example, young mice run around their cages more and, if you're an old mouse sewn to one, you benefit from an enforced exercise regimen. That means the advantages to an old mouse in a parabiotic pair go substantially beyond simply adding pro-growth signaling molecules or diluting bad ones.

These ambiguities didn't stop a surge of interest from scientists and Silicon Valley biohackers alike, who proceeded with varying degrees of scientific rigor. Continued parabiosis experiments have shown us that the older mouse in a heterochronic pair has improved growth of both brain cells and blood vessels in the brain, has better spinal cord regeneration, and can have an aged, oversized heart shrunk back to a more normal size. This extends the catalogue of organs which benefit from parabiosis and might harbour latent healing capacity, but doesn't bring us much closer to a workable therapy.

Others tried injecting youthful plasma into old mice and humans. There is some scientific rationale to trying this out just to see what happens: plasma transfusions are a relatively safe procedure, and a positive result would provide proof of principle which could then be built upon, rather like the original parabiosis experiments. However, human trials don't appear to have been a resounding success—one in South Korea hoping to use young plasma to alleviate frailty began in 2015 and has yet to report any results, and a U.S. trial gave transfusions of young plasma to Alzheimer's patients but didn't succeed in turning back the disease.

The field has also got something of a bad name thanks to private companies trying to cash in on the euphoria around young blood. One colorful outfit called Ambrosia offers anyone over 35 the opportunity to receive a liter of youthful plasma for $8,000 (at the time of writing, a promotional offer also allows you to get two for $12,000—buy one, get one half-price). In spite of reputed popularity among Bay Area tech execs and venture capitalists hoping to extend their time on Earth,* the company temporarily ceased treatments after the FDA (which regulates food and drugs in the U.S.) issued a statement warning that transfusions of young blood were risky and unproven. After spending nearly a year reassessing the rules, Ambrosia decided that their service is technically legal and resumed operations. The company also bills its treatments as a pay-to-participate trial, but there is no sign of any results at the time of writing. Worse, there isn't a control group against which to compare transfusion recipients, making it very difficult to discern any effects the treatment

---

* Rumors that Peter Thiel, a billionaire venture capitalist who co-founded PayPal, was interested in the procedure culminated in a storyline about anti-aging transfusions making it into the satirical sitcom **Silicon Valley.**

may have—you can't exactly give half your patients an infusion of saline solution if they've paid $8,000 for it, even if it would make for a fair test.

All this is despite a significant nail being hammered into the coffin for young blood theory in 2014, two years before Ambrosia was set up: a study giving mice regular injections of young plasma found that it didn't make them live any longer. This doesn't rule out some benefits for particular conditions—young plasma has subsequently been shown to improve liver function in old mice, for instance—but it suggests that the global effects of parabiosis can't be replicated by simple transfusions.

Meanwhile, the Conboys were working on wholesale blood-swapping between old and young mice, replacing parabiosis by connecting pairs of rodents to a tiny pumping device to exchange their blood. That is a pretty smart feat of mini-engineering in itself: mice have just one or two milliliters of blood,* so a microfluidic pump slurped out 150 **micro**liters at a time to allow it to be safely swapped between the old and young animals. After a few rounds of this back-and-forth pumping, the two mice have a pretty much 50:50 mixture of youthful and aged blood, and testing could begin.

This experiment is far less invasive than parabiosis

---

* By way of comparison, an average human has about five liters—several thousand times more, which is basically in proportion to the weight difference between us and mice.

and looks only at what's happening in the blood itself without prolonged sharing of organs. Even with a one-off exchange, the results were substantial and quite different from parabiosis. Young blood retained some of its rejuvenative powers, improving regeneration of muscle cells in the older mouse but, overall, the positive effects on the older mouse were outweighed by the negative effects of old blood on the younger one. Of the three tissues tested, muscle, liver and brain, the brain was the worst affected: not only didn't young blood rejuvenate brain cell growth in the old mouse, the older blood was clearly inhibiting growth of young brain cells, even though the tests were performed almost a week after blood had been swapped. Once again, the simple story of significant benefits from young blood seems to have been undermined—though there probably are some benefits, they're smaller than the negative effects of old blood.

Given that mass blood transfusions are both unlikely to work and impractical, how can we turn the results of these studies into therapies? The next step is to try to isolate which of the many facets of parabiosis is responsible for its effects. A few groups of scientists began trying to establish what changes from young blood to old and work out how to reverse it. This work involved cataloging molecular differences—what goes up, what goes down, what stays the same?—and then performing careful experiments to try to work out what their consequences are. One age-related miscreant identified is a protein called TGF-beta whose levels increase in old mice and old humans and which damps the activity of stem cells. By contrast, oxytocin—a hormone which has a complex role in behaviors from social bonding to sex and childbirth—is a potential beneficial factor in young blood which declines with age. A protein called GDF11 was also singled out as a youth-restoring factor, but subsequent work has thrown that finding into doubt. This kind of work has plenty further to run, because there are dozens of substances in blood whose levels change with age, and their good or bad effects may play out in combination with one another.

If this story is more about needing to modulate bad factors in old blood than needing to add regenerative youthful ones, one option is to adapt a treatment called plasmapheresis, a process similar to dialysis. In both procedures, blood is pumped out of a patient's body, has harmful substances removed, and is then pumped back in again, refreshed. Dialysis is used in cases of kidney failure to remove excess water and waste products from the blood which would normally be removed by healthy kidneys; plasmapheresis concentrates specifically on the plasma and is usually used to remove the antibodies which cause the immune system to go on a rampage in autoimmune conditions. If we can identify the problem molecules in old blood, we could reconfigure devices like these to remove them. The question here, which can only be answered empirically, is how often this treatment might need to be repeated: opting for plasmapheresis every few months would be a hassle, but perhaps an acceptable one if it substantially improved health; the punishing schedule of four-hour sessions three times a week endured by dialysis patients would be far less palatable.

The most straightforward approach is to try to optimize various signaling factors by altering their levels or effects with drugs. The Conboys tried dialing down the activity of TGF-beta, one of the signaling proteins they've identified as increasing with age, by giving mice a drug called an ALK5 inhibitor. (ALK5 is the receptor that cells use to detect and react to TGF-beta, so inhibiting it stops them from doing so.) The drug reawakened stem cells in the brain and muscle, causing new neurons to grow and speeding recovery of muscle after an injury. They also tried simultaneously administering the drug and extra oxytocin, whose concentration is reduced with age. This, too, had beneficial effects on brain, muscle and liver, very similar to those seen in heterochronic parabiosis, after just a week of treatment. The most exciting thing about this second study was that the addition of oxytocin allowed the dose of the ALK5 inhibitor to be reduced tenfold: from a practical standpoint, a lower dose of a drug reduces the risk of side effects in patients; from a theoretical one, this suggests that these signaling pathways interact in ways which means tweaking several at once can have an effect greater than the sum of its parts. ALK5 inhibitors and oxytocin are already approved for clinical use, meaning they're prime candidates for a first-generation signaling-correcting therapy in people.

Factors dissolved in blood are not the only contributors to the body-wide changes in signaling that accompany aging. Another key component of the signaling system is "exosomes"—tiny bubble-like packages which transport molecules between cells. The smallest of them are tens of nanometers across—hundreds of times smaller than a typical cell, and similar in scale to viruses. Their cargo varies, but often they carry messages encoded in microRNA—very short lengths of a molecule rather like DNA, which carries information as a series of bases (RNA uses the A, C and G we are familiar with from DNA, but with T replaced by U). When an exosome arrives at a destination cell it gets absorbed, depositing its cargo inside. There, the microRNAs can do their work, providing instructions which alter the recipient cell's behavior.

One study looked at stem cells in the hypothalamus—a part of the brain already heavily implicated in signaling to control fundamental processes like hunger, thirst, circadian rhythms and body temperature. The researchers found that hypothalamic stem cells died in droves as mice in the experiment aged. Injecting fresh stem cells from the hypothalami of newborn mice didn't just rejuvenate this particular region of the brain—it increased lifespan by 10 percent compared to mice given a different cell type as a control. And, as so often in aging experiments, the mice weren't just living longer, but living healthier, performing better in treadmill, muscle endurance and cognitive tests.

This is amazing: adding stem cells to just one place has effects so broad that the mice actually **live longer** as a result. While incredible, perhaps it's not surprising given the hypothalamus's role as a signaling nexus for so many diverse processes—adding stem cells should go on to add new neurons to this critical region, putting regulation of all the fundamental aspects of physiology under hypothalamic control back on track. However, the positive effects of injecting the stem cells manifested after just a few months—not long enough, by the scientists' estimations, for them to have got very far making new neurons. That meant that some more rapid process was responsible, leading them to suspect that signaling exosomes secreted by the stem cells were rejuvenating the cell population. Collecting exosomes from hypothalamus stem cells in a dish and injecting them alone, they saw many of the same anti-aging benefits.

If this result holds up, it could be turned into a treatment fairly directly: we could reprogram some cells into iPSCs and differentiate those into neural stem cells, which could be injected directly into the brain, or we could grow them in the lab and harvest the exosomes they produce for infusion. This probably isn't the only place where these tiny message-bearing capsules are significant in aging, nor the only place they could be used in therapy—one study showed that providing exosomes from neural stem cells dramatically improved recovery from stroke in pigs, and exosomes generally are under investigation as a way to deliver drugs and other useful molecules to where they're needed in the body. With exosomes, it seems that good things really do come in small packages.

Though the simple but alluring idea of young blood as a cure-all seems to have had a stake driven through its heart, the idea that aging is partly a phenomenon of signaling gone off the rails is taking wing. What all these experiments with heterochronic parabiosis, blood exchange, signal-tweaking drugs and exosomes show beyond doubt is that some aspects of aging and loss of regenerative capacity aren't just intrinsic to cells, but also reflect responses to signals in the cells' environment. What happens in aging is a vicious cycle: as the internal environment in our bodies worsens, cells and tissues affected by these aberrant signals deteriorate and then start to emit signals of their own which accelerate the body's decline. This is bad news in the spiraling decline of aging, where bad leads to worse; but it could be good news for us, as positive changes lead to virtuous cycles of rejuvenation in our bodies.

Whether we will be visiting a plasmapheresis

clinic every so often for a blood factor detox, taking drugs to rebalance our cellular signals or being infused with exosomes is yet to be determined—but fixing faltering signals is likely to be an important part of our anti-aging arsenal.

POWERING UP MITOCHONDRIA

The decline of mitochondria, the herd of semi-autonomous energy generators which can be found inside our cells, is responsible for aspects of aging around the body. There are fewer mitochondria in older cells, and those that remain are less effective at producing energy. The problems are especially acute in places where cells use a lot of energy like the brain, heart and muscles; mitochondria are almost certainly important in Parkinson's disease, and there's increasing evidence of their effects in other conditions too. Developing therapies to help out our mitochondria might thus alleviate many of the problems of old age.

As we mentioned in Chapter 4, the first theories about mitochondrial involvement in aging centered on free radicals. These are a family of voraciously reactive chemicals produced as a by-product of the high-energy reactions mitochondria use to generate power, and a particularly pernicious and mitochondria-relevant group is known as reactive oxygen species, or ROS. Unchecked, these oxygen radicals can go on a rampage around our cells, reacting with anything they see, damaging proteins, fats and even DNA. Luckily for us, they can be mopped up by antioxidants—molecules which can stabilize them without suffering significant damage themselves. Our bodies produce their own antioxidants, in the form of proteins like catalase and superoxide dismutase, and we also find them in our food in the form of vitamins, like vitamin C and vitamin E. Thus, if ROS are the problem, then there seems to be a simple solution: increase levels of antioxidants, either by telling our bodies to make more, or by taking vitamin supplements.

We also touched upon attempts to increase production of antioxidants in mice in Chapter 4: adding extra copies of antioxidant genes for superoxide dismutase and catalase doesn't seem to extend lifespan. In addition, there's a huge amount of evidence that antioxidant supplements don't increase lifespan, in mice or in people. A Cochrane systematic review (considered the gold-standard round-up of medical research results) published in 2012 examined 78 trials with a total of 300,000 participants to assess the effect of antioxidant supplements. The message was clear: these supplements are pointless, and possibly even harmful. Vitamins A and C along with selenium were found to have no effect on longevity; vitamin E and beta-carotene **increased** the chance of death by 3 and 5 percent, respectively.

The likely reason for the ineffectiveness of antioxidants is that ROS are used for a variety of functions around the body, as signals to send instructions within or between cells, or in more specialist applications like immune cells using them to destroy bacteria. As a result, mopping up too many free radicals with vitamin pills could cause our bodies to reduce production of our own internal antioxidant enzymes to make sure we've got enough ROS for these critical processes—or, at high doses, ROS levels could get too low to fulfill vital functions, actively causing harm. Thus, stopping these rampaging chemical species wholesale isn't an option.

However, there is one type of antioxidant which is still in contention for extending lifespan—those targeted specifically to mitochondria. Mitochondria, being the site of much of our body's ROS production, are also one of the main targets of their destructive streak, damaging their outer "skin" known as the mitochondrial membrane, their energy-producing machinery, and DNA (mitochondria, remember, have a short length of their own outside the main DNA storage found in our cell nuclei). This means they are at disproportionate risk from free radicals—and that protecting them from this damage could be disproportionately better for us than indiscriminately soaking up free radicals throughout the cell.

A 2005 paper reported results from mice which had been genetically modified with additional copies of the gene for antioxidant enzyme catalase, but altered so it would make its way to the mitochondria: they lived 20 percent longer than regular mice, with an average lifespan of 32 months rather than 27. Subsequent work in mice has shown that mitochondrially targeted catalase can reduce the risk of cancer in old mice, slow the progression of age-related heart problems, reduce the production of amyloid-beta and extend the lives of mouse models of Alzheimer's and improve muscle function in old mice.

There are also several mitochondrially targeted antioxidant drugs in the works. Probably the most advanced is MitoQ, which has made it as far as human clinical trials: one showed that it might help reduce inflammation of the liver in patients with hepatitis C, another that it improved blood vessel function in healthy people over 60, but a third showed that it didn't slow the progression of Parkinson's disease (though this might be because, as we've noted before, people showing symptoms of Parkinson's have already lost a huge fraction of their dopaminergic neurons). Interestingly, this might also be a good place to intervene to improve the function of telomeres: treating cells with MitoQ has been shown to decrease the rate at which they shorten. This implicates mitochondrial ROS damage to the telomeres' DNA in their shortening, as well as the number of times a cell has divided. Further clinical trials of MitoQ and other drugs like it are under way.

Another option is to ramp up our bodies' existing

capacity for mitochondrial quality control to try to get rid of the ineffective mitochondria and allow better-functioning alternatives to replicate and take their place. Drugs are being identified which can boost "mitophagy," the mitochondria-specific flavor of autophagy which we mentioned earlier. One compound is urolithin A, a molecule made by our gut bacteria while digesting nutrients found in our food, which has been shown to extend lifespan in worms, improve endurance and muscle strength in mice and slow cognitive decline in mouse models of Alzheimer's, as well as improve mitochondrial function in people over 60. Other contenders for mitophagy-boosting include spermidine, one of the DR mimetics we discussed a few chapters ago, and supplements that can increase levels of a molecule called NAD^+, which is critical to cellular energy generation, important to mitophagy and known to decrease with age.

However, it's quite possible that adding anti oxidants or improving mitophagy is essentially oiling a broken machine—something that will smooth the decline, but ultimately can't address its root cause. Unfortunately, finding the fundamental reason for mitochondrial decline with age is incredibly challenging. The more we learn about mitochondria, the more weird and wonderful these semi-autonomous symbiotic beasts in our cells seem, and the more complex their interactions grow. If there is a single root cause of mitochondrial decline, the strongest case is probably mutations to their DNA. Mitochondria, remember, are the only part of our cells outside of the nucleus that contain their own DNA, and increasing the number of mutations in it is enough to cause something that looks a bit like accelerated aging in mice. The question we really care about is the reverse: could reducing the burden of mitochondrial mutations slow down or reverse the aging process?

The best way to find out, as ever, is to fix it and see what happens. The most radical idea to prevent mitochondrial mutations is "allotopic expression": placing a backup copy of the mitochondrial genes in the cell nucleus with the rest of our DNA. While it may sound like an outlandish reengineering proposal, it's actually finishing a job that evolution started, but has never quite got round to completing. Mitochondria have a rather bizarre origin story: they are thought to have first come into being over one billion years ago, when an incredibly distant, single-celled ancestor of ours engulfed an entirely separate organism and began a symbiotic partnership that would last a million millennia. That engulfed mitochondrion-to-be had a full complement of DNA of its own but, in the intervening eons, almost all of its genes have either been lost or migrated to the cell nucleus. One reason this is a good idea is that it's a safer place to store DNA—where it will be away from all the nasty free radicals the mitochondria produce, be replicated far less often and protected by more efficient nuclear DNA repair mechanisms.

The idea of therapeutically moving mitochondrial genes to the nucleus isn't entirely new: it was actually first done in the 1980s, when yeast cells missing the mitochondrial **ATP8** gene were given a copy in the nucleus which was successfully imported into their mitochondria. Since then, it's moved on to become a treatment under serious consideration for inherited mitochondrial diseases, where mutations in mitochondrial DNA can cause problems ranging in severity from exhaustion during exercise to sudden death a few days after being born. The furthest through development is a therapy based on this idea for a mitochondrial eye disease called LHON. It's currently struggling in the final phase of clinical trials, but initial success in cells, mice and rabbits suggests that, even if this particular formulation doesn't work out in humans, there's at least something to it.

The idea of moving **all** the mitochondrial genes to the nucleus to combat mutations in aging, rather than just one which is causing a specific mitochondrial disease, came from Aubrey de Grey, who we met in Chapter 4 as the father of Strategies for Engineered Negligible Senescence, or SENS. Accordingly, it's his SENS Research Foundation that has got the furthest with this idea so far, successfully restoring function to cells in a dish with mitochondria missing two genes by providing them with a backup in the nucleus and, more recently, getting all 13 mitochondrially encoded genes working to varying degrees in the nucleus by optimizing their genetic code.

There's more work to do to get this working properly, and then to prove that it will actually help if we succeed. Perhaps the biggest outstanding question is: if this is such a great idea, why hasn't evolution already done it? Of around 1,500 genes needed to make a mitochondrion, over 99 percent are located in the nucleus in humans. Why did evolution stop there? On one hand, there's nothing particularly special about the number 15: a single-celled soil-dwelling creature called **Andalucia** has retained 38 protein-coding genes in its mitochondria, most of which are found in the nucleus in other organisms; **Plasmodium,** the parasite which causes malaria, has just three; and a parasite that infects oceanic algae was discovered in 2019 that may have no mitochondrial DNA at all (though there are differences in the details of how these organisms use their mitochondria to generate energy). There are also some significant roadblocks to evolution placing every last gene in the nucleus: at some point in our evolutionary history, the mitochondrial and nuclear genomes diverged to have slightly different "dialects," meaning that, nowadays, if some mitochondrial DNA just found itself in the nucleus by accident it would require significant and extremely unlikely levels of modification to make the necessary protein.* If this is why our mitochondria still have DNA, there's not really any theoretical barrier to moving it, just a large wall of biological improbability stopping it from happening by chance.

However, it might be that mitochondria need to retain this small complement of genes in order to work efficiently: mitochondrial DNA may represent devolved government, using local knowledge to optimize cellular metabolism somewhat independent of the centralized bureaucracy of the nucleus. If this is the case, a backup copy elsewhere could destabilize cells' delicately delegated metabolic chain of command. The only way to know for sure is to try it: at a minimum, we'll learn a bit of mitochondrial biology; it's quite possible we'll find cures for some mitochondrial diseases along the way; and, at best, we eradicate mitochondrial mutations as a cause of degenerative aging.

The final option is to take on not the defunct mitochondria, but the cells they take over. When mitochondrial DNA is altered, most of the mutations don't cause significant problems—but a small percentage of cells come to be dominated by zombie

---

* For example, in mitochondria, the three-letter DNA sequence TGA means "add the amino acid tryptophan," whereas you might remember from Chapter 3 that TGA is the nuclear DNA code for "stop reading." This is a pretty fundamental obstacle, as it would mean a protein would stop being built midway, rendering it functionally useless.

mitochondrial clones which have lost a significant chunk of their DNA, and with it their capacity to generate energy. It's not entirely clear if cells like this are sufficient in number to cause significant age-related problems, but, as we've seen with senescent cells, it's quite possible for a few bad apples to accelerate aging. That suggests a similar approach to how we deal with senescent cells: kill them, but with (currently hypothetical) "mitolytic" rather than "senolytic" drugs. The risks of this idea are much the same as the risks of senolytics, in that removing the delinquent cells might cause its own problems. For example, a mitolytic drug targeting a muscle fiber containing defective mitochondria would destroy the whole fiber, leading to muscle wasting— exactly the type of strength-sapping process we're trying to avert. However, it's probably worth trying: the worst-case scenario is that we'll come away with a better understanding of what these cells' internal zombie apocalypses mean in aging, and the best is that the trade-offs are worth it, and killing these cells benefits our health.

Overall, we've got quite a few options when it comes to trying to slow or reverse the contribution of mitochondria to aging, but we're not entirely sure which will work best. This is partly because we still don't have a complete picture of what happens to our mitochondria as we get older. In the short term, treatments might include mitochondrially targeted antioxidants to mop up the free radicals they produce, or supplements, like urolithin A, to ramp up our bodies' own quality-control mechanisms. In the long run, it might be possible to reengineer our biology to make sure that mitochondrial mutations no longer matter, eradicating their contribution to degenerative aging—a goal well worth expending significant energy to achieve.

REPELLING THE ATTACK OF THE CLONES

Damage to our DNA, and the mutations that result from it, could be some of the hardest age-related damage we need to fix in our bodies. The first approach we could take is probably the most obvious: repair. The second relies on understanding exactly how mutations cause problems in our aging bodies, and learning how recent developments in DNA sequencing could overturn old ideas about why mutations matter.

The first, obvious approach would be to improve our DNA repair machinery. Our cells have gone to incredible lengths to ensure that damage to our DNA can be fixed—as we noted in Chapter 4, an average cell is thought to take up to 100,000 hits to its DNA daily, so if even a tiny fraction of those were to stick it could be catastrophic. There's a mind-blowing array of DNA repair processes involving hundreds of different genes to spot problems, cry for help and excise any damage, showing us without doubt that this is of serious concern to our bodies.

However, as we know from our understanding of the evolution of aging, even something as important as repairing DNA will only be as good as it needs to be to allow us to pass on our genes.

This means that inspiration can be drawn from the animal kingdom, where many animals seem to be more mutation-proof than we are. Take the bowhead whale, whose exceptional longevity we remarked upon in Chapter 2. As well as living for over two centuries, these graceful giants can weigh up to 100 tons. In spite of this immense body size, bowhead whale cells are about the same size as cells from a human, or even a mouse; thus, given that they weigh over 1,000 times more than a typical human, they have approximately 1,000 times as many cells as we do. This means they have very roughly 1,000 times as many opportunities for a cell to acquire the mutations that would give rise to cancer, and two or three times longer lifespans for them to do so (more if you consider human lifespans in the wild). In spite of these handicaps, our marine megafauna aren't riddled with hundreds of tumors. This general rule—that larger animals with more cells and often impressive longevity tend not to succumb to astronomical levels of cancer—was first formulated in 1977 by medical statistician Richard Peto and is called "Peto's paradox."

Peto's observation seems to be a paradox only between species, not within them: there's evidence that taller people are at greater risk of cancer than shorter ones, and the larger breeds of dog are more cancer-prone than smaller ones, too. In these cases, more cells but the same species-specific cancer defenses means a greater likelihood of cancer overall. (Don't panic, tall people: the stats suggest that you're at less risk of cardiovascular problems and dementia than short people, so the difference in overall mortality probably comes out in the wash.) This further bolsters the idea that we might learn something from large, long-lived animals: it suggests that mere size doesn't confer big animals with some unknown cancer-protective advantage.

Recent sequencing of the elephant and bowhead whale genomes provide tantalizing hints about how we might go about improving our own mutation resistance. The elephant genome contains twenty copies of a gene called **p53,** while we humans only have one. **p53** is the gene most often mutated in cancer, and has been dubbed "the guardian of the genome," thanks to its crucial protective role. It's a gene with many functions, one of which is to cause apoptosis or senescence in cells whose DNA is badly damaged. Thus, it could be that these additional copies of the gene make elephant cells particularly prone to precautionary suicide, making cancer less likely to develop. Bowhead whales don't have extra **p53,** but do have additional copies of or subtle variations in genes responsible for DNA repair, which may make mutations less likely in the first place—there's more than one way to skin the cat of cancer prevention.

Applying these findings naïvely is a risky business: while removing a copy of **p53** from mice does indeed make them dramatically more prone to cancer, adding an extra one causes symptoms of accelerated aging and reduces their lifespan. It's hypothesized that the trigger-happy suicide protein causes too many stem cells to die, making the mice cancer-proof but causing them to run out of stem cells prematurely. I certainly won't be lining up for gene therapy to provide me with an extra copy of **p53** given these results, and it's likely that evolution has made complicated trade-offs between degeneration and cancer prevention that can't be understood simply by counting numbers of genes. As biologist Leslie Orgel famously noted, evolution is cleverer than you are.

However, this approach is not without hope: you might remember from earlier in this chapter that mice given telomerase plus three DNA-protective genes lived longer than normal ones. One of those protective genes was **p53.** As we also saw in Chapter 2, evolution hasn't optimized for longevity, but reproductive success. In the case of these genetically modified mice, perhaps extra **p53** causes more cell death, but plenty of other cells with longer telomeres can divide a few more times to cover for their lost peers. Evolution might avoid such an approach because extending telomeres and making extra cells

takes a lot of energy, and having the normal complement of both **p53** and telomerase already postpones
both cancer and stem cell exhaustion to a point long
after most mice in the wild are dead. That means
we don't necessarily have to be cleverer than evolution to reap improvements in lifespan: inserting
additional copies of single genes might be too simplistic an approach, but it could be that adding or
altering a handful of genes, done judiciously, can
achieve positive results long before we have a full
understanding of how the many protective systems
in our cells interact. (We'll be talking more about
gene editing for longevity in the next chapter.)

Improving DNA repair is an approach that
is agnostic toward how the damaged DNA and
resulting mutations actually go on to affect our
bodies: it slows down the accumulation of damage so
its consequences should be delayed, too. However,
the next class of approaches we could take to save us
from mutations depends critically on our emerging
understanding of how mutations affect our tissues
in old age.

The first suggestions that mutations could contribute to the aging process were made in the late
1950s, just a few years after the double-helical structure of DNA was discovered in 1953. The idea was
that our genetic code would accumulate random
mistakes throughout life; since this code gives the
instructions to build proteins, these errors would
lead to changes in the structure of these proteins;

and, as we've seen, proteins' structure dictates their function, so these changes would mean that our cellular components would gradually get less effective with age, quite possibly increasing levels of DNA damage in the process, leading to a vicious cycle which drives aging.

Modern DNA sequencing technology leads us to question this simple picture. We now know how many mutations cells accumulate over a lifetime—data that scientists even a decade ago could only really guess at—and the numbers don't quite add up. This shows us that most cells in your body acquire around ten to fifty mutations every year you're alive. The uniformity across tissues is surprising—whether it's a cell in the lining of your intestines, constantly dividing and besieged by toxins in food, or a brain cell in its cosseted environment which might not divide for your whole life, almost every cell type we've checked falls into this relatively narrow range.

One of a handful of exceptions is sun-exposed skin:* cells from these areas can accrue ten times as many mutations per year. Indeed, skin researchers (nearly) equate skin aging with sun exposure—how much ultraviolet light a given patch of skin has received over its lifetime is a significant predictor of how biologically old it is. Another place

---

* The comparison to non-sun-exposed skin often makes use of skin from the buttocks, presumably making frequent nude sunbathers difficult subjects for study.

where mutations abound with atypical frequency is the lining of your lungs if you're a smoker, for obvious reasons.

These mutation rates mean that a 65-year-old might expect to have a couple of thousand mutations in any given cell in their body, and maybe 10,000 in a cell of sun-exposed skin or a smoker's lung. That sounds like a lot, but it might not be enough to cause widespread problems with proteins: only just over 1 percent of our DNA codes for proteins,* and cells only use those specific ones that are relevant to their particular function, meaning that the chance of any one mutation falling in a protein-coding region important to a given cell is low; some mutations, known as "synonymous" mutations, won't make a difference to the protein anyway (this is because there are a number of ways to "spell" each amino acid in the DNA code, so changing a letter won't make a difference if the resulting "word" codes for the same amino acid); and, finally, because most genes come in two copies—one from each parent—even if a mutation does arise in one, there's a backup which can usually step in and fill the gap. If you work through the math, the number

---

* The other 99 percent of our DNA used to be called "junk DNA," but we now know that's something of a misnomer—it's involved in various processes like making sure the right proteins are produced at the right times. However, in most cases its exact DNA sequence is less critical than the protein-coding regions, and the odd mutation is less important.

of cells which would expect to have problems in both copies of a protein important enough to cause issues is very small—maybe one in every few thousand of our cells. Perhaps mutations aren't such a big deal after all?

It would come as a significant relief if widespread random mutations aren't one of the reasons our cells deteriorate with age, because it would be almost impossible to fix. If every cell contained dozens of different mutations which were functionally significant, we'd need to find some technology which could go in and correct the errors, and it's very hard to envisage what that might be. Even if you imagine creating an artificial mutation-repairing nanobot, it would need to carry a reference copy of your whole genome in its back pocket against which to check every single possible error, and dreaming up such a sci-fi solution would push any hopes of curing aging back into the twenty-third century.

However, though detailed sequencing of aged genomes makes this scenario seem less likely, it has revealed a more subtle way in which mutations can affect the aging process. The best way to understand what's happening is to look at the most famous consequence of mutations in our DNA: cancer.

Fundamentally, cancer is a disease of accumulated mutations. A cancerous cell needs to acquire specific faults in its genome in order to turn into a tumor: perhaps most important, it needs to disable genes which cause its cells to stop growing, enable genes which encourage growth or both; a key part of this, as we've seen earlier in this chapter, is to activate telomerase or, occasionally, some other mechanism; as the cancer develops, it will need mutations to grow its own blood supply and suppress the immune system; and, at some stage, most cancers turn off critical DNA repair mechanisms, allowing genetic chaos which enables more mutations, which makes all of these tasks easier. The same math that says you'd expect very few cells to have even a single crippling mutation in old age tells you that accumulating enough specifically pro-cancerous ones in the same cell is also extremely unlikely in current human lifetimes.

Unfortunately, cancer has an ace up its sleeve: evolution. When a non-cancerous cell acquires what scientists call its first "driver" mutation, the key feature is that it provides the cell with an evolutionary advantage. What an evolutionary advantage means for a cell is the ability to outgrow neighboring cells with unmutated DNA. Imagine that a normal cell obtains a mutation which disables a gene whose purpose is to stop its growth under particular conditions: it could then start dividing, even though all of the adjacent cells think (correctly) that there's no need. It might make a few thousand, or even a million, daughters before some other still intact growth-modulating process kicks in—our own whole-organism evolution has equipped us with multiple, highly redundant mechanisms for most cellular functions, not least to reduce the odds of cancer—and this period of rapid growth is brought to a halt. This process of rapid but temporary growth is a "clonal expansion," so named because it's an expansion of cells which are clones of one another, sharing the same driver mutation.

This process of driver mutation followed by clonal expansion makes cancer dramatically more likely. There are now thousands or millions of cells which already contain one driver, making it far more likely that one of them could get "lucky" and acquire a second driver mutation. This second mutation can then cause another clonal expansion, meaning there are now a million cells with **two** pre-cancerous mutations, ready and waiting for a third . . . and so on. This is evolution by natural selection, but operating on the cells within organisms: cells with a growth advantage multiply, outcompeting their obedient, rule-following neighbors, repeatedly making the next step on the path to cancer more likely.

Driver mutations are central to our modern understanding of how cancer arises. A millionfold increase in odds at every stage is what makes cancer possible within a human lifespan. In fact, around half of us are predicted to be diagnosed with cancer given current life expectancies. As a result, cancer is a major killer: it's responsible for over a quarter of deaths in the rich world, and one in six worldwide.

Thus, cancer is the first reason to be concerned about accumulation of mutations with aging—

anything we can do to decrease their frequency should help us prevent cancers from arising in the first place. However, the process of clonal expansion is problematic in another way—it also provides a new mechanism for troublesome mutations to have an outsized effect on our aging bodies, even if they only occur in a comparatively tiny number of cells.

We're just beginning to find out how widespread these clonal expansions are. Recent studies have shown that, in some tissues in old age, there's hardly a normal cell to be found. But it's not a mosaic of a million mutations, every cell different; it's a patchwork of little colonies, usually under a millimeter across, each made up of cells with one or two specific mutations that give them a competitive advantage. This was first uncovered in a 2015 study looking at the skin of four people over the age of 50, which found that 20–30 percent of cells contained a driver mutation—an average square centimeter of skin played host to 140 different driver mutations. This is a mind-blowing result: all over your skin, right now, are thousands of competing clones vying for dominance. Given this sheer number of clones and the vast burden of mutations, it's only the action of protective mechanisms like cell death, senescence and the immune system that keep us all from very rapidly succumbing to cancer all over our sun-kissed skin.

However, sun-exposed skin was always thought to be an outlier given its constant bombardment from ultraviolet light. Subsequent work looking at the esophagus (the tube connecting your mouth to your stomach, and accordingly well shielded from sunlight) found strikingly similar results. Though there were far fewer mutations overall, individual clones were able to expand further. (It's likely that the decreased ability of clones to expand in the skin is one of the mechanisms which protects us from skin cancer, but we don't know exactly what confers this protection.) By the time you're elderly, your esophagus will be home to around 10,000 different clones covering almost its entire interior.

The easiest way to understand why this might be a problem is to look at an example. Another place where clonal expansions are common is among the HSCs—the stem cells responsible for making blood cells. The most common driver mutation in HSCs is a gene called **DNMT3A.** The protein it encodes controls whether a stem cell divides asymmetrically (forming one stem cell and one differentiating daughter) or symmetrically (forming two stem cells). An HSC with a **DNMT3A** mutation will preferentially divide symmetrically, which provides a huge competitive advantage—rather than just producing one stem cell after each division and therefore keeping numbers constant, you get two, then four, then eight, doubling each time, and, after just 20 divisions, it will outnumber a puny asymmetrically dividing stem cell by a million to one. It's no wonder that these cells outcompete those with intact **DNMT3A** and clonally expand.

The mutants do eventually stop their rampage because other mechanisms kick in telling them that that's enough dividing for now. However, someone with this mutation can nonetheless end up with a large fraction of their HSCs being mutants, which in turn means that many of their red and white blood cells are manufactured by these mutant clones. The presence of clonally expanded HSCs is associated with the risk of cancers like leukemia—so far, so obvious, given that cancer is made possible by a succession of clonal expansions—but is also associated with diabetes, and a **doubling** of the risk of a heart attack or stroke.

The precise mechanisms behind these observations are not yet fully understood, but it's plausible to expect things to go wrong in your blood if a significant number of your blood cells have a mutation in a key gene. Scientists studying this have observed that people with clones in their HSC pool have less consistently sized red blood cells, which, at a minimum, is a sign that something's up. We also know that the atherosclerotic plaques behind heart disease and some strokes are mostly composed of white blood cells, specifically macrophages—it would make sense that dysfunctional mutated macrophages could worsen their prognosis. Similar clonal expansions around the body in rapidly renewing tissues from skin to gut linings could well cause subtle problems of their own that accelerate specific diseases or aspects of the aging process if left unchecked.

Thus, these new studies have breathed new life into the threat of mutations as a factor in aging, by the same process of evolution and clonal expansion which makes cancer a menace. The key principle, in both cancer and aging, is that what's good for the survival of an individual cell isn't necessarily good for the organism as a whole. Multicellular organisms rely on collaboration between cells, and these selfish runaway replicants don't perform their functions quite as they should, decreasing the fitness of the mouse, human, or whatever creature they find themselves in.

The reason for that relatively long digression is that this new understanding of the effects of mutations should inform attempts to devise treatments. The first point to note is that it's not a done deal that these clonal expansions are sufficient to bring about disease and dysfunction, nor that random changes in individual cells that don't lead to clonal expansions are entirely irrelevant. Step one is to gather more data, and this effort is already under way: huge reductions in the cost of DNA sequencing mean that scientists are rushing to measure more tissues in more people in more detail than ever before. The next ten years will see our knowledge transformed, revealing what mutations occur where in far greater detail, and that will allow us to work out where and how they might cause problems. But if, as seems likely, these clonal expansions are an issue, what can we do?

The first piece of good news provided by this picture of how mutations contribute to aging is that, though the clones are very widespread, the takeover is usually orchestrated by cells with defects in only a handful of genes. Mutations that inactivate a gene called **NOTCH1** are responsible for the majority of the clones in both skin and esophagus, so a drug which targeted this mutation could significantly reduce the number of aberrant cells we'd need to worry about. Going after the top five or ten mutations would dramatically improve matters, which is far better than the thousands of treatments that the random-mutations-causing-dysfunction theory would have required us to develop. Exactly what those treatments would look like is speculative for now, but there are plenty of options: cancer researchers have spent decades looking for "targeted" therapies which will attack cancer cells with particular mutations while leaving bystander normal body cells alone, meaning that there are many treatments which could be repurposed to go after non-cancerous clonally expanded mutants.

As well as tackling clonal expansions relevant to aging, this could also be a good preventative medication for cancer. A study examining the mutations present in over 2,500 tumors found that half of the earliest driver mutations occurred in just nine genes, and that they appear years or perhaps even decades before diagnosis. That means, in principle, that a substantial number of cancers could be prevented if we could find ways to kill cells with one of those nine mutations. While again we don't know exactly how we would do that, finding treatments targeting nine genes is a substantially more tractable problem than if hundreds or thousands of genes bore responsibility for setting a cell on the path to becoming a tumor.

The other way we could overthrow the clones rather than just killing them is to change the environment to favor normal cells. In evolution, "survival of the fittest" specifically means fittest **for the current environment.** Thus, if drugs or other therapies change that environment such that regular, unmutated cells have a competitive advantage, they would be able to gradually wrest control from the clones. How exactly we'd do this is again up for grabs, but there is preliminary evidence that the tables can be turned on the mutants. A recent proof-of-concept study found that giving mice a dose of X-rays boosted growth of cells in their esophaguses with **p53** mutations (**p53**, as well as being a favorite mutation in cancer, is also the second-most-common driver of clonal expansions in the esophagus), but that zapping them with X-rays **after a dose of antioxidants** improved the odds of the normal cells competing with them. While therapeutic antioxidants-plus-X-rays might not catch on as a treatment, this work does demonstrate that the supremacy of clonal expansions is contingent on their environment, and that tweaking it can help normal cells regain the upper hand. (It also raises the intriguing prospect that perhaps the reason too much radiation exposure causes cancer isn't mainly because the radiation directly causes mutations, but that it might encourage mutant clones to grow bigger, increasing the odds of one of them taking the next step toward cancer.)

Perhaps the most radical idea to obviate DNA damage and mutations is a complete refresh of the body's stem cells—an approach which works whether it's clones or random mutations in individual cells (or both) that cause problems in aging. The other piece of good news from sequencing studies is that there are usually still at least a few cells left over whose DNA doesn't have any significant problems. If we could extract some of those unmutated cells, turn them into induced pluripotent stem cells, double-check their DNA is free from random errors or game-changing driver mutations and then turn those iPSCs into stem cells for skin, esophagus, intestine, blood, and other tissues plagued by mutations, we could use these to replace the mutant cells. More knowledge will also inform where in our bodies a DNA refresh is most needed, and it might well be that fixing the most urgent tissues tides us over until we can fix those where mutation rates are lower, or clonal expansions less problematic.

Mutations in our DNA will probably be one of

the hardest hallmarks of aging to overcome. The ideas we've discussed range from proofs of concept in the lab to purely speculative for now. But it's a problem we're going to have to deal with eventually: mutations will keep accumulating, 10 to 50 per year in every cell in your body, and clones will keep expanding, slowly strangling their normal neighbors, even if we sort out absolutely everything else. The happiest solution to this problem would be to discover through further study that non-cancer-causing mutations aren't sufficiently harmful to our health to be a significant problem even in extended lifespans, but we should plan for the worst even while we hope for the best.

Thankfully, our understanding of the prevalence and type of mutations in aging will increase dramatically in the near future. Genome sequencing is now cheaper than ever, and interest in accumulation of mutations is coming not only from the aging research community, but from the cancer research community, too. Those of us interested in curing aging need to make sure that this work explores the contribution of mutations to degeneration as well as cancer, and ensure that attempts to actually treat it are developed in tandem with exploratory work. If we do that, the first treatments to tame rampant clones or gene therapies to bolster our DNA's defenses could be near enough to matter for many of us alive today.

8

Reprogramming Aging

Having removed, replaced and repaired what we can, the final stage in an actual cure for biological aging will almost certainly require us to reprogram our own biology, hacking what nature has given us to prevent problematic processes from happening in the first place. Since the "program" for our biology is written in our genes, this will involve editing them to optimize the good stuff, reduce the bad and add new capabilities to our cells and organs.

While this might sound futuristic, there's plenty we can do in the foreseeable future of medical care. It won't be long before we can use gene editing to optimize the hand that evolution has dealt us, and we could even perform cellular reprogramming—another name for the process of creating induced pluripotent stem cells—to turn back the clock not just in cells in a dish, but in our whole bodies.

These ideas provide us with a glimpse of the final step in biogerontology, and perhaps all of medicine: combining everything we've learned so far into intricate computer models of human biology, which will make everything we've discussed seem primitive. Once we've achieved this goal, we will be truly ageless—and, as "aging" gradually loses its meaning, we'll probably cease to call the treatments we devise "anti-aging" at all.

UPGRADING OUR GENES

Our DNA is the blueprint for our bodies, from their large-scale layout to the tiniest components which govern interactions within and between cells. As we discover more about the human genome, there seem to be ever more stories in the press about a "gene for" this or that, and it can be tempting to lapse into "genetic determinism"—believing that your entire biological future, risk of disease, lifespan and even much of your personality, is fixed by the contents of your genetic code.

Life is, of course, more complicated than that. It's obvious that DNA does affect lifespan: after all, humans can live for more than 100 years, but a nematode worm can only live for a few weeks, and that difference is encoded in our DNA. We've also seen that mutations in single genes can dramatically change the lifespan of worms and mice in the lab. Can we put this knowledge to practical use to extend the healthy lifespan of people?

The first question is to what extent our genes determine our longevity normally. Clever studies with identical and non-identical twins and analyses of large cohorts allow us to estimate the degree of "heritability" of longevity—the extent to which long or short life can be inherited. The size of this effect turns out to be surprisingly small—somewhere around 25 percent. However, more recent work has revised down even this low estimate. Unfortunately for statisticians, people don't usually pick their partner entirely at random—they tend to pair up with people whose characteristics are more similar to themselves than would be expected by chance, a tendency called "assortative mating." A 2018 study, using thousands of birth and death records from a genealogy website, mathematically corrected for this effect—and found that the heritability of longevity dropped to under 10 percent. The researchers actually found that married couples' lifespans were more closely correlated than the lifespans of opposite-gendered children.

This is empowering news for many of us: your lifespan isn't written in your DNA, and you needn't see your parents' longevity as a ceiling on how long you can hope to live. With the right diet, exercise, lifestyle and a bit of luck, our destiny is in our own hands to a larger extent than simplistic genetic determinism would have us believe.

However, it's disempowering news for any biologists hoping to comb through the population looking for the genetic underpinnings of longevity: genetic effects are subtle and we shouldn't expect to uncover amazing longevity mutations if we go out naïvely looking in the general population without carefully correcting for issues like assortative mating. Fortunately, the task can be made a lot easier if we look in more unusual places. The first place to try is the exceptionally old.

There's something distinctly strange (in a good way) about people who make it to 100. Studies find that they weigh about the same and don't smoke or drink much less, exercise much more or eat much better than the general population. In spite of this, they not only live longer, but they seem to put off getting age-related diseases until later, too. A study of U.S. centenarians found that they spent dramatically less of their lives with disease: 9 percent for centenarians versus 18 percent in the wider population. They also retain their independence for longer, with the average centenarian in the study still able to perform everyday tasks until their 100th birthday.

As we move from the merely normally old to the exceptionally old, the heritability of longevity seems to increase. Whether your mom or dad lived to 70 or 80 doesn't mean a great deal in terms of your own likely lifespan, but if one of your parents made it to 100, or even beyond, it might be time to sit up and pay attention. You may have noticed this yourself—perhaps a friend's family, or, if you're very lucky, your own, can trace in its family tree a dynasty of long-lived women (statistically, they usually are women: female centenarians outnumber males by over five to one). This is one anecdotal observation which does stand up to statistical scrutiny: if one of your siblings lives to 100, you stand about a ten times greater chance than someone in the general population of doing the same.

As a result, geneticists have gone fishing for versions of genes which are overrepresented in people who have made it to 100. This has proved to be a surprisingly tough gig, but two genes do come up pretty consistently: **APOE** and **FOXO3.**

The **APOE** gene codes for a protein called Apo-E which is responsible for transporting cholesterol around your body, and which version you have has a huge effect on your odds of both cardiovascular problems and dementia. It comes in three variants—**APOE2, APOE3** and **APOE4.** The most common variant is **E3**—about two-thirds of people globally have the plain vanilla **E3/E3** genotype (i.e., one copy of **E3** from your mom, and one from your dad), and that gives them a 20 percent chance of developing dementia during their life. The **E4** variant is less common, but it's bad news: having one copy, present in about 25 percent of people, increases the risk of Alzheimer's to almost 50:50. People with two copies (thankfully just 2 percent of the population) are almost certain to develop Alzheimer's, and are diagnosed at an average age of 68—about a decade earlier than those with **E3/E3** genes. **E2**, on the other hand, seems to be protective, roughly halving the lifetime risk of dementia if you have a copy from one parent—and maybe reducing risk by a further factor of four if you have two copies. It's a similar story with heart disease, with **E4** carriers again at higher risk.

No surprise then that your **APOE** genes can have a significant effect on your chance of making it to 100. The **E4** variant is significantly underrepresented in centenarians compared to the general population— because many of those carrying **APOE4** died of heart disease, dementia and so on before making it to their 100th birthday. It's not a guaranteed death sentence—there are a handful of centenarians even with the unluckiest **E4/E4** genotype—but, if you're going for a longevity record, it's an additional hurdle to clear. Studies suggest that those with two **E4** genes live slightly fewer years on average than those with **E3/E3**, while having two **E2** genes can extend your life by a little.

The next most promising longevity gene is **FOXO3.** Not only are variants of it associated with extreme longevity in human centenarians, but there's also evidence for its importance from model organisms. Thanks to evolutionary conservation, we share many of the same genes with organisms as distant as flies and nematode worms. **FOXO3** is very similar to the worm gene **daf-16**—which, like **daf-2** and **age-1** that you'll remember from Chapter 3, affects lifespan in worms via the insulin signaling pathway. Like their worm counterparts, humans with a favorable **FOXO3** variant probably experience a mild genetic simulation of dietary restriction, enjoying effects like increased autophagy which slow their aging down just enough for these flavors of **FOXO3** to be noticeably more common in people who make it to extreme old age.

The other unusual place to look for longevity genes is isolated populations. Imagine you were carrying a hotshot longevity mutation which would add five years to your lifespan. When you died at 91, rather than the 86 you'd have made it to without the mutation, it would be unlikely to come to the attention of the scientific or medical communities—91 is a good run, sure, but it's far from unheard of. If you have a couple of children, you might pass it on to one of them, and they might pass it on to one of their children, and so on. Unless the mutation makes your descendants have more children on average, it won't spread through the population, but just drift, with any increase or decrease in frequency down to chance alone.

In isolated communities, however, mutations can persist. If there's not a huge population in which to dilute them, they can come to be present in a substantial fraction of the much smaller population purely by chance. It could spread into different families and, a few generations later, two people who are descendants of the original carrier could meet, fall in love and have children of their own. This phenomenon is what underlies the danger of having children with siblings or cousins—if both parents carry a rare "recessive" version of a disease-causing gene (which is fine if you only carry one copy), there's a one-in-four chance that their son or daughter will carry two copies, and develop the health problem that mutation causes.

This is exactly the chain of events that led a three-year-old girl, part of the Old Order Amish community in Berne, Indiana, to the hospital in the mid-1980s—a trip which would eventually lead to the discovery of a new gene for human longevity. After bumping her head, the girl developed a large pool of blood under her scalp, and surgery to drain it only made things worse—she nearly bled to death. A few years later, surgery for a dental abscess nearly caused her to drown in her own blood. There's a collection of "bleeding disorders" which could be responsible for blood not clotting to seal up a wound but, one by one, doctors eliminated them. This girl's condition was something that, at that time, was unknown to medical science. It was only due to the persistence of a doctor and blood-clotting specialist named Amy Shapiro that the underlying cause of her bleeding was uncovered.

Trawling the literature for clues, Shapiro read about a protein called PAI-1 which is involved in blood clotting. She eventually managed to persuade a colleague to test the sequence of the gene that codes for it, **SERPINE1,** in the girl's DNA. They found a two-letter mistake—the DNA-copying process had stuttered, turning a TA into a TATA. This tiny change meant that she was totally lacking working PAI-1, causing her blood-clotting issue, but she seemed entirely normal otherwise. Further testing showed that her parents had one copy of the mutated **SERPINE1** each, meaning that they produced less PAI-1 than usual—and they seemed to be entirely unaffected, with even their blood clotting in the usual way.

Other research had found that people with higher levels of PAI-1, caused by a different mutation, were at higher risk of cardiovascular disease. This leads to a natural question: if more PAI-1 is worse, is less better? The Amish community in Berne provided an ideal test case, and Shapiro applied to the National Institutes of Health for funding to perform a study. Her grant application was rejected: the NIH didn't think 100 subjects would be enough to discern a statistically robust effect. How wrong they were. In 2015, almost two hundred Amish people volunteered for a battery of medical tests examining their blood and heart health. Those with a single copy of the mutated **SERPINE1** gene had slightly better cardiovascular health than those with two copies of the regular gene along, interestingly, with longer telomeres. They were also much less likely to get diabetes—eight of the 127 non-carriers had diabetes, against zero of the 43 with the mutation. Most strikingly, using genetic testing and family trees to deduce the genotype of relatives who had already died, the study found that people with the mutation lived **ten years** longer on average than those with two copies of the normal gene, boosting average lifespan from 75 to 85 years.

How is this possible? Well, in the intervening decades since its discovery we've learned that PAI-1 is a protein which isn't just involved in blood clotting; like most genes, it is involved in a number of different processes around the body. Perhaps most crucially when it comes to aging, PAI-1 is associated with cellular senescence: it's both involved in a cell's internal decision-making when it considers becoming senescent and a component of the senescence-associated secretory phenotype by which senescent cells wreak their havoc around the body. Dialing down the odds of senescence and the potency of the SASP are both plausible ways to extend lifespan. While the carriers of a single copy of the mutation seem not to have any problems with blood clotting, it's also possible that a slightly less eager clotting system might be of benefit as you get older, reducing the odds of problems like stroke.

We should still be a bit cautious about PAI-1 reduction. Though the effect is impressive, it was found in a small population, and the result could still be a fluke, or something highly Amish-specific. Regardless, this surprising lifespan increase should cause us to revisit our expectations about how large an effect single genes can have on longevity in humans. As we saw in Chapter 3, those 1970s evolutionary biologists who thought that mutations in single genes could necessarily have only very small effects on lifespan were quite wrong—and PAI-1 proves that not only were they wrong in worms, but also in people.

As well as the Amish, there are many other isolated populations worth studying. We have already met the Ecuadorians with a growth hormone receptor mutation leading to Laron syndrome (which combines short stature with substantial reductions in cancer and diabetes). Ashkenazi Jews have also been the subject of much study, and have shown that less extreme growth-hormone-related mutations than those behind Laron syndrome are also linked to long life. A woman in a huge extended family in Colombia, many of whom are predisposed to get Alzheimer's in their forties, made headlines in 2019 because she managed to evade the condition for decades longer than her relatives—seemingly because of an exceptionally rare mutation in both of her **APOE** genes. Populations with a restricted gene pool will undoubtedly continue to provide intriguing results for both biogerontologists and scientists studying many other aspects of human biology.

Our search for longevity-boosting genes can also be informed by an understanding of what the genes actually do based on work in the lab. There's a long list of genes that, when either disabled or provided in additional copies, extend the lives of model organisms like worms, flies and mice. We've already met a few of these, such as **age-1** and telomerase, but there are many more to choose from. For example, mice given an extra copy of a gene called **Atg5** display increased levels of autophagy, and live 17 percent longer; there are multiple genes related to growth hormone, like the mutation found in longevity record-holding Laron mice; and a gene called **FGF21** which simulates dietary restriction, and can extend mouse lifespan by a third.

So, having scoured the genomes of model organisms, isolated communities and the extremely old for hints as to how we could live longer, how might we make use of this knowledge? The traditional way is to develop drugs to mimic the effect of a beneficial genetic change. In the case of PAI-1, for example, it seems from the tale of the Old Order Amish that we don't need as much of it in our body as we normally have. Thus, scientists are looking for molecules which inhibit PAI-1, molecules which will gum up the protein and stop it carrying out its functions. One drug now under development improves diabetes, blood cholesterol and fatty liver disease in overweight mice, and has passed preliminary safety tests in people.

Creating drugs which interfere with the action of particular proteins is the classic way to implement findings in genetics, and has been the basis of many medical breakthroughs over the last few decades. However, there is also a more radical approach: gene therapy. Gene therapy is the idea that we could go in and tweak our DNA—directly add new genes, remove ones we don't want, or replace defective ones with better alternatives. Gene therapy can be more permanent than drugs: if the DNA is "integrated" into your genome, it will remain there indefinitely, and there's no need to take the medicine every day. It also has the potential to reduce side effects: drugs will often have "off-target" effects, interfering with proteins or processes they're not meant to in addition to the intended ones; a gene therapy for a single gene will, by definition, only affect the gene itself—and, though single gene changes can of course cause wider knock-on effects, it may well be less than a drug that hits multiple proteins and pathways simultaneously.

Unfortunately, gene therapy in adult organisms is difficult. The first problem is getting the new genes and editing machinery into potentially trillions of cells. We don't yet have the tools to reliably edit every cell in the human body, which means that, if your proposed edit relies on universality, you're in trouble. The most common "vectors" for inserting DNA are viruses, whose modus operandi is to insert their own genetic information into our cells to produce copies of themselves. If you strip out the viral genes and replace them with a gene you'd like to insert, the virus will obligingly deliver that instead. However, our immune systems are always on the lookout for invading viruses and can sometimes overreact: it was a disproportionate immune response to the viral vector, rather than the gene therapy itself, which caused the death of Jesse Gelsinger in 1999. Jesse was an 18-year-old who received one of the first experimental gene therapies, but died four days later. This tragedy caused a massive reputational setback to the field. There's also the risk of the wrong piece of DNA being altered and, as always, the risk of cancer if some aspect of the DNA editing goes awry.

However, huge strides are being made in gene editing both because it's very useful for scientists doing experiments in the lab and because it has huge therapeutic potential. A technology called CRISPR has made headlines—and indeed earned the 2020 Nobel Prize in Chemistry for co-discoverers Emmanuelle Charpentier and Jennifer Doudna—because it makes gene editing more precise and far cheaper, and has already found its way into human trials to treat disease. It has been limited to making modifications outside of the body for now, meaning that cells can be safety-tested before being put back into patients. There's also a buzz around adeno-associated virus (or AAV)—which was used to deliver telomerase to adult mice in the previous chapter—thanks to its ability to evade the immune system, and to provide a DNA payload which doesn't "integrate" into the genome, reducing the risk of cancer. There are a handful of approved AAV therapies, and hundreds more in human trials.

A study published in 2019 took the first step in using AAV gene therapy to treat multiple age-related diseases in adult mice. Trying three different genes identified in studies of aging, both individually and in combination, the most successful formulation combined a gene which reduces levels of TGF-beta, one of the bad factors identified in old blood we discussed in the previous chapter, and **FGF21,** which we mentioned is a gene which mimics DR. Mice given this dual gene therapy lost weight whether they were young and fed a high-fat diet or obese due to older age, saw a reduction in diabetes and recovered better when kidney or heart failure were induced.

The study authors have founded a company in the U.S. called Rejuvenate Bio with the intention of commercializing combination gene therapies against aging, and the next step is to try them in dogs—in particular, Cavalier King Charles spaniels which suffer a high rate of age-related heart problems. If trials are successful, the plan is to get the therapy up and running in pets, aided by simplified regulatory approval, and their shorter lifespan allowing results to be obtained more quickly. The market for these treatments for animals is expected to be worth billions of dollars, and revenues could then be used to develop human versions.

As well as adding extra copies of beneficial genes, there's also the possibility of reducing our load of detrimental ones. One example is **PCSK9,** a gene responsible for controlling the amount of cholesterol in our blood. A study in Dallas, Texas, found in 2005 that some African Americans had very low levels of LDL "bad" cholesterol, caused by a mutation which disabled the gene. Further research found that this mutation—which occurs in about 3 percent of African Americans but less than one in 1,000 Americans of European descent—resulted in a staggering 88 percent reduction in the risk of heart disease. This led to a race to develop drugs to reduce its activity in the rest of us, and "PCSK9 inhibitors" are now considered the gold standard in cholesterol-reducing medication, used in people with high cholesterol that can't be controlled by statins (including people with one of several **PCSK9** mutations that increase rather than decrease its activity). Trials have begun with **PCSK9** "RNA interference" technology, which intercepts the RNA molecules that are the intermediary allowing DNA to be turned into proteins; this could reduce levels of PCSK9 for months at a time with a single dose. If these can reduce cholesterol and heart disease risk without problematic side effects, the next step could be to disable the gene entirely. This has already been shown to work in mice using CRISPR, and a company called Verve

Therapeutics is developing a version of this therapy for human trials.

Finally, there's scope for modifying our existing genes to optimize them for longevity. One option for small changes is a modified version of CRISPR known as "base editing," which allows changing of a single DNA letter at a specific point in the genome. One promising target for this therapy in the not-too-distant future would be **APOE**: the **E3** variant differs from both **E2** and **E4** by just a single DNA letter. We also already know that different variants can coexist in the same person without disastrous consequences and, even better, **E4** is bad (and **E2** good) in a "dose-dependent" manner—two copies is worse (or better) than one, which is worse (or better) than zero. Thus, even if you didn't manage to edit every single copy of **APOE,** it's quite likely that you'd still see a positive effect.

If we're to realize the power of gene therapy for aging, we need to do more work to understand how these genes work individually and in combination. The 2019 study treating age-related diseases in mice also looked at a third gene, **Klotho,** an extra copy of which can extend mouse lifespan by about 25 percent (it takes its name from Clotho, one of the Three Fates in Greek mythology who spun, measured and chopped threads to determine the length of human lives). However, using all three genes together actually reduced the effectiveness of the treatment: the study found that **Klotho** and **FGF21** don't play nicely together. In biology, the whole is often not the same as the sum of the parts—it can be either greater or, as in this case, lesser, for reasons which are rarely obvious at the outset.

In the long run, gene therapy is likely to have a huge role in medicine. The possibility of more targeted treatments which don't depend on people remembering to take tablets every day is significant enough, before we consider the ability to modify human biology by adding extra health-giving features which aren't available in our natural genome. The good news for those of us hoping to treat aging with gene therapy is that the field as a whole continues to move very quickly, with new clinical trials being announced at a very rapid pace. As is now a familiar story, they will start out being used in patients at serious risk of disease—such as using CRISPR to modify **PCSK9** in patients with such high cholesterol that they are at risk of a heart attack in their thirties or forties. If those patients avoid side effects, including the risk of cancer thanks to modifications gone wrong, we could gradually see their use being expanded to cover people with less serious health problems, like age- or diet-related high cholesterol, and eventually maybe all of us will receive **PCSK9**-modifying therapy as a preventative "vaccine" for high cholesterol.

Looking to the further future, gene therapy could be used to radically reengineer human biology. We've seen in previous chapters how it might help to give our cells copies of genes which aren't present in any human's DNA, such as novel enzymes to break down undegradable waste in our lysosomes, or backup copies of at-risk mitochondrial genes. Even these are primitive compared to what we could eventually hope to do with genes, genuinely reprogramming our biology. We could create entirely new genetic "circuits" that, rather than just pumping out a protein, can respond to changes in our bodies and stabilize our biology in the face of the destabilizing effects of aging. Though our lifespans are not heavily determined by our natural genetics at the moment, gene modifications reprogramming our cells' capabilities will probably be a significant component in how we ultimately go about curing aging. We'll discuss this in far greater detail in the final part of this chapter. First, we'll examine the radical effect of just four genes which have the power to reverse aging in cells . . . and maybe whole bodies, too.

TURNING BACK THE EPIGENETIC CLOCK

Throughout this book we've learned that the process of aging is surprisingly malleable. Whether it's dietary restriction, genetic changes or sewing a mouse to a younger one, the rate of aging can be slowed; with senolytic drugs, telomerase and other therapies in development, we may be able to reverse it. This is incredibly exciting news, and I hope it has by now changed your view on both aging and medicine. But perhaps the plasticity of aging shouldn't be so surprising. Aging is a solved problem, after all: even if their parents are old, babies are born young.

No matter whether her parents are in their teens or their forties, a baby is born at age zero, with brand-new organs and skin as smooth as . . . well, a baby's. Babies inherit their parents' DNA, but not their chronological age. This is a key part of disposable soma theory from Chapter 2—while our bodies are expendable, the "germline" cells involved in reproduction can't be if your species is to survive. The germline is immortal: the fact that you're reading this means that your parents, their parents before them, and an uninterrupted and unfathomably lengthy lineage of great-great-great- . . . great-grandparents, right back to single-celled organisms on the early Earth, all successfully had children—and they must have been biologically young enough to have children of their own. The successful preservation of the germline over several billion years isn't **technically** immortality, but it's not a bad start.

On one level, this is maddening. We all have the tools to build an entirely new, fresh life written in our DNA, and yet we clearly don't have the tools to perform what seems like the far simpler task of keeping what's already built running. It's all very well that mother nature can do this with newborns, but can we uncover her tools and make use of them in medicine?

We've actually already discussed one of the methods by which science could make this exciting idea possible. In Chapter 6, we met the process of deriving induced pluripotent stem cells, or iPSCs, the incredible multitalented precursor cells which we can make from regular differentiated body cells. Since this process was discovered, we've found out that inducing pluripotency seems to rejuvenate cells in a way that mimics the magic tricks nature uses to grant babies their youth. The process of making iPSCs is termed "reprogramming" the cells, and thus this idea is known as rejuvenation by reprogramming.

The first line of evidence for this rejuvenation is the epigenetic clock, the freakishly accurate predictor of biological age based on epigenetic marks on your DNA we met in Chapter 4. Steve Horvath actually found this in the 2013 paper where he announced his clock. Having established that it worked over many different types of tissue, he made one final test of its predictive power: he used it to calculate the epigenetic age of both embryonic stem cells—"naturally" young cells isolated from a human embryo just a few days after sperm met egg—and iPSCs, derived from cells obtained from adults. The embryonic cells had an epigenetic age close to zero, which makes sense. The adult cells used to make the iPSCs had a normal epigenetic age corresponding to that of their donor, which also makes sense. But the iPSCs themselves were epigenetically zero years old—their biological clock had been reset,

making them indistinguishable from their embryonic counterparts.

Experiments since have doubled down on this
finding: fully functional iPSCs have been successfully derived from people as old as 114, and the cells
have an epigenetic age of zero whether the donor
was a young adult or a centenarian. Even better,
differentiating these iPSCs into specific cell types
leaves their epigenetic youthfulness intact. This
means that you can take a 90-year-old's skin cells,
make iPSCs and differentiate them back into skin
cells again—and those new skin cells will themselves
be young. This is already brilliant news: it could up
the effectiveness of all the stem cell therapies we've
got planned if we can use iPSCs as the source of the
donor cells—the new brain cells, eye cells, blood
stem cells or whatever we produced from them
would be young again, ready for another few decades of use and abuse.

Even better, it appears that the epigenetic reset
doesn't occur in isolation but is accompanied by
other rejuvenative effects. The iPSCs also have
better-looking mitochondria, and lower levels of
mitochondrial reactive oxygen species. They also
have longer telomeres, comparable to those found
in embryonic stem cells. This is all almost suspiciously great news: by inserting extra copies of just
four genes—the Yamanaka factors, O, K, S and
M, the discovery of which earned Yamanaka the
Nobel—we seem to be able to reactivate a molecular

deep-cleaning process analogous to that which undoes the ravages of time in the germline.

There are some caveats: for example, there's a faint epigenetic shadow which allows iPSCs made from young and old donors to be distinguished, though it seems to fade if you let the iPSCs divide a few times. However, even if the details are still being ironed out, the process of inducing pluripotency seems to reliably reverse the aging process in cells. That's exciting—but can you do it in whole animals?

The first piece of good news is something we mentioned in Chapter 6: you can inject iPSCs into a mouse embryo and generate a fully functional mouse. This is pretty solid evidence that they behave exactly as regular embryonic cells do—and, in particular, that they're not prematurely aged somehow in a way that stops the newborn mice from functioning properly or causes them to die young. We can also examine the lifespan of cloned animals. Dolly the sheep was cloned from her "mother" by taking the nucleus from an adult cell and then inserting it into an egg whose nucleus had been destroyed. Her birth led to speculation about life for clones in general—born from old DNA, even transplanted into a young egg, could she have a normal life, and lifespan?

Six and a half years later, the answer seemed clear: Dolly had to be put to sleep after she developed a cough, and a follow-up X-ray revealed multiple tumors on her lungs. This is an unusually brief life for a Finn-Dorset sheep, a breed which often survives to nine or more. She was also diagnosed with arthritis when she was five—again, unusually early—and a measurement of her telomeres when she was one year old had shown that they were shorter than those of other young sheep. All this together led scientists to suspect that starting with an adult cell nucleus from a six-year-old mother may have meant that Dolly started life with a biological handicap, eventually leading to premature aging and a stunted lifespan.

However, subsequent work has overturned this supposition. A more thorough study looking at 13 cloned sheep, four of whom were genetically identical sisters of Dolly cloned from the same cells, found that they were all aging pretty normally in their dotage. Aged between seven and nine at the time of the study, a thorough ovine medical found their cardiovascular health, blood test results and joints to be similar to non-cloned sheep of the same vintage. It might well be that Dolly just got unlucky—aged four, she'd caught a respiratory virus known to cause lung cancer which was sweeping through the flock at the Roslin Institute, where she lived. It's pretty likely that this was the cause of her tumors, meaning that they weren't a sign of premature old age after all. Given this simple explanation, which was known at the time of her death, her persistent citation as evidence that clones die young seems rather peculiar.

Experiments with mice have gone further, first cloning a mouse, and then taking a nucleus from a cell of the clone and putting it into an egg cell again to make a clone of the clone . . . and so on. Early studies seemed to show that the success rate dropped with each "generation," so cloning a clone was harder than simply cloning a "normal" mouse had been, and cloning the clone of a clone was harder still. One meticulous attempt to understand this eventually made it to the sixth generation, which took an incredibly frustrating 1,000 attempts to get a single live baby mouse—only for its foster mother to promptly eat it, ending the experiment. It's hard to imagine what it must be like to meticulously inject 1,000 nuclei into egg cells, **finally** have it implant successfully in a mouse womb, wait out the pregnancy . . . and then for Mom to chow down this tiny piece of scientific history before you've had chance to run any tests.

However, cloning techniques have since improved significantly, and this decreasing efficiency with each generation has disappeared. A group overseen by the same scientist who witnessed that appalling act of mouse cannibalism published a paper in 2013 documenting the success of repeated cloning over 25 generations with no obvious increase in difficulty over time. Most crucially from our point of view, the great-great- . . . great-grandclones were healthy and had normal lifespans. Once again, the magic of cellular reprogramming resets the aging clock with each generation. (The experiment continues and, at the time of writing, is on generation 43.)

The final piece of evidence for rejuvenation by reprogramming hails not from the lab, but the ocean. The jellyfish **Turritopsis dohrnii** is a half-centimeter, 90-tentacled sea creature also known as (spoiler alert) the "immortal jellyfish." Its immortality is granted by a two-way life cycle, where adult jellyfish (the "medusa" phase) can do a Benjamin Button and age backward to the junior, "polyp" stage. This seems to happen by a process of cellular dedifferentiation. The polyp can then grow up all over again, becoming a betentacled medusa once more, ready to repeat the process of aging backward whenever life's cumulative stresses get too much for it. This jelly-phoenix shows us that having a baby isn't the only way that biology has found to turn back the clock—it's possible to do it to a complete adult body of aged cells, too.

There's an obvious problem with us humans adopting the jellyfish's strategy. Reprogramming cells wholesale in a live human would result in cells in all our vital organs—lungs, heart, liver, kidneys—losing their function and turning into pluripotent stem cells. Stem cells might be powerful in terms of their potential, but they're hopeless in practical roles, like pumping blood round your body. Generating undifferentiated cells willy-nilly would result in catastrophic organ failure and rapid death. The other risk, if the organ failure didn't get you, would be total wipeout from teratomas—the disgusting matted-hair-eyes-and-teeth tumors formed by pluripotent cells. Even a single iPSC in a live organism can be deadly, so purposely inducing them body-wide would be catastrophic.

This isn't just a grim theoretical prediction—we've tried it in mice, with exactly those results. Multiple cancers and organ failure blighted two experiments attempting to make **in vivo** iPSCs published in 2013 and 2014. However, a couple of years later, a more subtle approach enjoyed greater success. Scientists took mice genetically engineered to have both a premature aging disorder and extra copies of the Yamanaka genes in their cells which would only be turned on if they were taking a particular drug. After letting them develop normally and begin to show signs of premature aging, the drug was administered. If administered continuously, the poor mice only lasted a few days before being overwhelmed by organ failure—this turned on the genes continuously, and was effectively a repeat of previous attempts. Lest this sounds like a needlessly cruel rerun of an experiment with a known outcome, the scientists did have a plan: this time, they were searching for a safe duration of Yamanaka factor activation which would turn back the aging process a little without sending the mice full jellyfish. When they reduced the dose to two days on, five days off, things looked very different. This cyclic activation of the Yamanaka factors improved their heart function, speeded recovery of both muscle and pancreas from injury, made them look younger and added 30 percent to their lifespan overall.

This was very much a proof-of-principle experiment: as we've discussed before, mice suffering from premature aging are not the ideal model for this kind of study because it might be easier to fix what you broke than to perform the same feat in the slow-motion chaos of normal aging. Nonetheless, it's an exciting result: this ridiculous-sounding treatment, activating a bunch of genes discovered for an entirely different purpose almost a decade before, seems to do something important to the aging process.

The excitement generated by this preliminary finding means that follow-up work is starting to fill in the gaps. We've shown that this isn't a mechanism confined to mice. Human cells in a dish subjected to transient activation of the Yamanaka factors and a couple of other genes show an across-the-board turning back of their biological clocks but, crucially, without losing their cellular identities. The process knocked a few years off their epigenetic clocks, pepped up their mitochondria, increased levels of autophagy . . . the only thing that wasn't changed was telomere length, which is probably actually a positive, because it means that the cells hadn't been reprogramed into iPSCs with active telomerase. This work also showed that human muscle stem cells extracted from the body, transiently reprogramed and then injected into mice can help aged muscles regenerate. Transient reprogramming has been shown to improve healing of eye injuries in middle-aged mice. We also now know that reprogramming can be delivered and activated safely in adult mice over the medium term: five-month-old mice given OKS (without the M) gene therapy have survived for over a year without obvious ill effects. Things are moving so fast with reprogramming that there will probably be some new findings by the time you read this.

These results are encouraging, but far too preliminary for you to go down to your doctor and demand a prescription for OKSM (quite apart from the fact that they'd look at you quizzically because there's no way to deliver them to humans yet). The key to turning this into treatments will be to unpick what's actually happening as we induce pluripotency, and in what order. Experiments on cells in a dish show that it's a multistep process: the first stage seems to be scrubbing off the epigenetic signs of old age, and it's only after that has been largely completed that the cell begins its dedifferentiation journey proper, from adult cell to stem cell. It needn't be this way—the process of epigenetic clock reversal could easily be simultaneous with that of dedifferentiation, or it might be that a cell has to get all the way to being an iPSC cell before it begins its spring cleaning. Luckily for us, the Yamanaka factors' path to pluripotency seems to be ordered favorably for those proof-of-principle experiments in mice we have just looked at: it means that two days' worth can turn back the aging clock, but stopping the drug after that means their cells don't have long enough to dedifferentiate and kill the mice in multiple unpleasant ways.

The fact that these things happen sequentially rather than all at once also suggests that they are at least somewhat independent, which means we could envisage genes or drugs which affect a cell's aging without affecting what type of cell it is. An ideal outcome could be a pill which resets all the age-related changes, but which leaves it as whatever type of cell it was originally, doing its good work in the body.

We can actually already do the opposite—changing a cell's identity without changing its biological age— by a process called "transdifferentiation," or direct reprogramming. This procedure is very much like generating iPSCs, but a different cocktail of genes turns one type of body cell straight into another— turning a skin cell straight into a neuron, for example—without the iPSC step in between. This could be very useful medically, allowing doctors to turn cells from one plentiful type to another which is needed in the body without going via iPSCs and putting the patient at risk of cancer. This is under investigation to generate new insulin-producing cells from other pancreatic cells for patients with diabetes, as well as new heart muscle cells and new neurons. These have the potential to be useful thera-pies but, more interestingly for us in this context, being able to alter a cell's identity without changing how old it is provides another piece of evidence that these two things are addressable independently.

There are several ideas to get this working in humans under active investigation. The classic approach would be to try to find a drug which mimics the effects of OKSM, or awakens those genes from their slumber in differentiated cells in the body. Not only is this the traditional way to unleash the power of lab discoveries as medicine, it has the significant advantage of not requiring us to inject a clutch of very powerful genes into our cells. You can always stop taking a drug; it's much harder to undo changes to your genetic code. Of particular concern is the M in OKSM, also known as **c-Myc,** because it's an "oncogene" often aberrantly activated in cancer. There are several approaches to "chemically induced reprogramming" under investigation in the lab, and scientists have successfully made iPSCs, neural stem cells and neurons from adult cells without the need to insert any genes. As we learn more about them, it's possible that these compounds could be turned into anti-aging drugs.

There's also a lot of interest in variations on gene therapy. For those justifiably cautious about **c-Myc,** the good news is that reprogramming doesn't seem to need it—as we mentioned a few paragraphs ago, OKS seems to work too, and has been safely used in mice for over a year. Researchers are trying to separate out the effects of O, K, S and M to identify which are active at which stages in reprogramming, what they each do and whether we could get away with even fewer of them. There's also the opposite approach, using extra genes which have been shown to improve the efficiency of reprogramming of cells in a dish: researchers are adding to the alphabet soup with combinations like OKSMLN (L for **LIN28,** and N for **NANOG,** a gene used by embryonic stem cells whose name derives from Tír na nÓg, the land of eternal youth in Irish and Scottish mythology) to find out what their effects are in animals.

There's even a case for ditching OKSM altogether. Since the discovery of induced pluripotency, iPSCs have been totemic; creating these godlike cells which are able to form any other type of cell has been the abiding goal of much stem cell research. However, iPSCs aren't necessarily the most useful target—there's actually no need for cells that versatile given that what we ultimately care about is working adult body cells. Perhaps we shouldn't rely on Yamanaka's factors, but search for new genes which turn back the clock to a less primordial stage in cellular development. One suggested timepoint is the so-called embryonic–fetal transition which, in humans, occurs around the eight-week mark during pregnancy. Before it, any injuries to the developing baby heal flawlessly; afterward, wounds heal imperfectly, leaving scars like those we're all familiar with after a cut or scrape. Turning back the clock to this stage of life could give our cells improved regenerative powers but, proponents of this idea hope, without the risk of accidentally overshooting into pluripotency, cancer and chaos.

Finally, there are options which step away from the idea of trying to emulate reprogramming. The fact that the first part of reprogramming seems to be the reversal of age-related epigenetic changes adds weight to the idea that epigenetics is a causal mechanism behind the aging process, rather than simply a clock face giving us a readout of our age. If this is the case, perhaps we'd be better off steering clear of the black magic of the Yamanaka factors and focusing on reprogramming our epigenetics directly. There are now modified versions of CRISPR that can alter epigenetic marks at multiple locations in our DNA simultaneously—and scientists are working on technologies to edit hundreds, or even thousands, of sites. That would mean we could think about precision approaches to replicate what OKSM does by brute force. However, the black magic approach retains a certain appeal—if we can let nature's own tools restore epigenetic order in our cells, we might be able to avoid the troublesome process of working out what exactly needs changing in order to do it ourselves.

Exactly how we'll disentangle the different kinds of epigenetic and other changes which occur during the wide-ranging process of induced pluripotency and transdifferentiation is yet to be seen. There are undoubtedly years of hard scientific graft ahead of us before we fully unpick the details, but what gives me hope that we might see therapies in years or decades rather than centuries is how easy it seems to have been to stumble upon these phenomena almost by accident. Yamanaka wasn't seeking a fountain of cellular youth when he found his four factors and generated the world's first iPSCs: he was specifically looking for something which would revive their ability to differentiate into anything they liked. Our good fortune has been that his OKSM factors seem to do both. But for this serendipitous success, it would be very hard to convince a skeptical scientist that we could turn back the cellular clock by activating just a few genes. The fact that it works is inspiration to go out in search of genes or drugs specifically for their powers to turn back the epigenetic clock, revitalize mitochondria, extend telomeres and so on, which the Yamanaka factors have shown us how to do as a side effect.

Therapies based on careful use of these reprogramming factors, or clever drugs or other treatments which can mimic their effects, may not be so far away. They may even come sooner than some of the otherwise more straightforward therapies discussed in previous chapters, partly due to the sudden intense interest in therapeutic reprogramming thanks to these promising early results. Using transient reprogramming to turn back the aging clock is one of the most exciting ideas in biogerontology: it sounds absolutely crazy at first, but the evidence so far suggests it might just work.

REPROGRAMMING BIOLOGY AND CURING AGING

Aging is a phenomenally complex process. Nonetheless, as we've seen in the last few chapters, we have good ideas as to how we might treat it. All these ideas have, at a minimum, precedent in the lab, and most of them aren't just speculative treatments based on theory or experiments on cells in a dish.

If we could get several, most or even all of these treatments working as preventative medicines for people, it would be a massive achievement. It would be very likely to significantly improve health in old age, and certain to make a huge contribution to our understanding of which are the most significant contributors to aging, and how these different phenomena interact. However, impressive though this would be, I don't think it would cure aging on its own.

You'll have noticed throughout this book that many of the changes associated with aging are connected: senescent cells have wide-ranging effects thanks to their pro-inflammatory SASP, impacting signaling, the immune system and cancer risk, all while being caused by things we've met elsewhere— short telomeres, DNA damage and mutations— which we can also imagine treating directly; stem cells crop up repeatedly, as treatments, as something which we could use signals to fix in old age and, just now, as the inspiration behind cellular reprogramming; chronic inflammation is both a cause and an effect of cellular senescence, immune aging, and so on. The hallmarks of aging are the names of key interchanges on a biological network more like a map of the New York Subway than a bulleted list— and it will take some work to establish the precise route of every line, and identify every station.

To truly cure aging, we need to take a more holistic, "systems biology" approach. We need to understand that our cells and bodies aren't made up of a collection of isolated phenomena, each of which can be fixed one at a time, but a complex system of components interacting in tangled networks with each other and even themselves.

The therapeutic ideas we've discussed tackle individual hallmarks of the aging process, eliminating one kind of cell or returning something that changes with age to more youthful levels. Even if they are beneficial overall, these treatments are very likely to have side effects on the other aspects of our biology. Maybe senolytic drugs will make us live longer, but our first-generation attempts to target them will be slightly overzealous in their cell removal and could eventually result in stem cell exhaustion. We might be able to compensate by adding more stem cells, or by using telomerase, altered signals or epigenetic reprogramming to encourage existing ones to divide a few more times, which might put our mitochondria out of kilter, or do something weird to our kidneys or our brains. All treatments have unintended consequences, as doctors are all too aware.

We are going to have to learn to reprogram human biology. As we begin to understand how its different components interact, we will gradually find more intelligent ways to intervene. Our biology involves interactions of molecules within individual cells, within and between whole populations of cells, with the extracellular matrix they sit on, with the immune system, the brain, our genes, our environment, and so on. Tweaking one part of this system will send ripples out into the others. We need to make sure that those ripples stabilize the system overall: we can't just focus narrowly on success measured against whatever they were designed to target directly.

The other reason we need to approach human biology in this holistic way is that individual humans are quite different. Mice in a lab are often genetically identical and raised in exactly the same environment as one another, meaning that any two mice will have a far more similar profile of age-related changes than any two humans. We'll need more advanced ways to measure whether your genetics, lifestyle, environment or just luck have predisposed you to, say, mitochondrial mutations in your lungs while leaving you with lower-than-average levels of sugar-modified collagen in your arteries, for example, and give whatever treatments we devise for those in proportion to your own personal spectrum of age-related changes, how you will react to the treatments, and whether you'll be particularly susceptible to any given side effect.

Cellular reprogramming is a glimpse of what a systems approach to treating aging could look like, albeit a very simple one which we've happened across almost by accident. The four Yamanaka factors are so named because they're "transcription factors," which is a biological term meaning that they are genes whose function is to influence the behavior of many other genes. They are not workers on the factory floor, but high-level managers whose invocation has far-reaching cellular implications— as it would have to, given that you can entirely reset a cell's identity by summoning them. The result is that just four genes can perform a monumentally complex task whose details we still don't fully understand, all by using existing biological circuitry inside cells.

The Yamanaka factors were found by trial and error, but, if we understood this cellular circuitry, we'd be in a position to rewire it in a far more purposeful way. If we could understand how circuits inside cells send and receive signals between cells, our reprogramming would become smarter as we began to untangle the consequences of the changes we're making around the whole body. Once we can integrate this broader knowledge with the ways age-related changes affect these systems, we'd be in a position to develop smart treatments with maximum benefit and the minimum of side effects.

Perhaps we'd activate two of the Yamanaka factors in the liver, heart and intestines, reduce the activity of three entirely different genes in some kinds of cell in the brain and add a new gene we'd custom-built for a particular task in the immune system; maybe later we could insert a small package of artificial DNA with its own programmable logic, doing x if it's in a cell where y is high and, if not, z and a in proportion to levels of something else. These kinds of programs are how our cells already work, with transcription factors regulating which other genes are turned on and off depending on the environment, signals and other transcription factors, so it's not implausible—even if gaining the understanding and technological prowess to do it ourselves will be challenging. If we truly master a systems-based approach to biology, it may well be very hard to describe the complex therapies we devise except in some kind of systems biology programming language: narrative language isn't designed to express the ludicrous complexity of emergent phenomena arising among vast numbers of interacting actors, but mathematics is. The increasing mathematization of biology will expand our ability to describe and then predict its complexity in ways that words simply cannot capture.

Once we have the models which would allow us

to predict these outcomes, biology will be transformed. Initial research will no longer be conducted **in vitro** (literally "in glass"—meaning cells in a dish or molecules in a test tube) or **in vivo** (in living things like worms, flies and mice) but **in silico**: in a computer. We are already taking the first small steps toward **in silico** biomedicine, and advanced models and simulations will eventually allow us to test all kinds of theories far more quickly and reproducibly than messy lab biology, and only the most promising therapies will need to be evaluated in slow, expensive trials in mice or people.

If this sounds futuristic, that's because it is. We are just starting to understand how networks of genes interact inside cells, and how signals are sent around the body. We're some way off building detailed, predictive models for human biology. That said, it's important to put the potential timescales into perspective. Even if you think the first actionable predictions of these computer models may be 50 years away, it's still important to lay the groundwork now: 50 years is still soon enough to benefit billions of people alive today, especially if we can add years to healthy life expectancy with some of the therapies already on the drawing board. And, though a fully fledged model of all of human biology could be even further away than that, it's also quite likely that the first attempts which, though imperfect, build and improve upon our current ways of doing medicine will come sooner.

In 2012, scientists created a computer model which could simulate a bacterium called **Mycoplasma genitalium.** This, as the name suggests, is a sexually transmitted bug which also holds the title of smallest known self-replicating bacterium. As a single cell with just 525 genes (we humans have around 20,000), this is as simple an organism as you can model—but, as well as explaining existing experimental observations, the model was able to predict behaviors that had never been observed before which were then corroborated in the lab. It's small but it's a start, proving the principle that computers can simulate biological systems. The **C. elegans** simulation we mentioned in Chapter 3 is perhaps the next step, though moving from one cell in a bacterium, to 959 cells in a worm, to tens of trillions in humans will undoubtedly be a challenge.

There are some examples of simple computational and systems medicine approaches being used in humans. One example is HIV treatment, where mathematical models allowed scientists to establish how fast different stages of HIV's life cycle progressed—and, once these models uncovered how fast the virus replicates and mutates, it became clear that multiple drugs could be used simultaneously to stop the virus from rapidly evolving resistance to single treatments. Though a cure for HIV remains elusive, modern combination therapies inspired by this insight can keep the number of viruses in patients sufficiently low that they can lead relatively normal lives, including safely having sex without a condom without putting their partner at risk of infection. There are other examples at an earlier stage where researchers have started using machine learning models to uncover new uses for existing drugs by looking at which proteins they affect, the drugs' molecular structures and so on, and predicting other uses they could be put to, either individually or in combination. One recent study used this approach to train a computer model to recognize the characteristics of a list of known DR mimetics, then used it to identify other drugs which might have similar life-extending effects.

The technology underlying these models is growing at an exponential rate. First, our ability to collect the kinds of data we need is growing incredibly quickly. Genome sequencing is the poster child of biological data collection, and it's getting rapidly cheaper: in 2001, just after the completion of the Human Genome Project, sequencing a human genome cost about $100,000,000; by 2008, the price had dropped a hundredfold to $1,000,000; in 2019, a whole-genome sequence cost under $1,000. Genome sequencing and related techniques are "omics" technologies. These techniques are said to be "unbiased" because you don't have to choose in advance exactly what you're looking for: rather than sequencing a single gene thought to be involved in a process, or measuring levels of a particular protein as we would have done in the

past, we can look across the whole genome (with genomics) or all the proteins in a given population of
cells (proteomics), and so on. This offers far greater
opportunities to find the unexpected, and to see
how cells and organisms behave as interconnected
biological systems.

We also have an exponentially increasing ability to process this kind of data. Computing power
has doubled every two years since the 1960s, as famously observed in Gordon Moore's eponymous
law. A similar, albeit less smooth, trend has given us
dramatic improvements in computer storage at an
even faster rate. It would be a mistake to extrapolate these trends into the indefinite future, as we are
likely to run into physical limits relatively soon —
the last half-century of processing power improvements has been driven by making the components
on microchips ever smaller, and we are approaching
the minimum size permitted by the laws of physics.
But we should be able to continue advances in data-
crunching speed by making algorithms more efficient, making chips optimized for particular tasks
like machine learning, and using new technologies like quantum computing.

Though past performance is no guarantee of
future success, trends like these make it seem quite
plausible that we'll have both the vast amounts of
data and the computational chops to process it that
we'd need to create detailed models of human biology. Given how far we've come in the last 50 years,

it would be foolish to bet that the kind of systems biology we'd need to cure aging won't be possible in the next 50.

The idea of cellular reprogramming is captivating. It makes me oscillate wildly between wondering if we just might have got incredibly lucky and, thanks to Yamanaka, been handed a cheat code for cellular biology, or whether its apparent success in the lab is a cruel joke played by nature, presaging a string of frustrating failures to turn it into practical treatments. However, even though it's not really a systems biology approach, it shows us the way: a seemingly sideways intervention in a bunch of genes which have no obvious role in aging, but whose combination allows a substantial reversal of the arrow of time.

Regardless of whether this first iteration of reprogramming results in a useful therapy, I believe that our entire approach to biomedicine will ultimately be best described as reprogramming: we'll need to quantify and utilize the interactions between the uncountable elements of our biology, add new features where our own genes don't already have the tools, and do all of this in a programmatic way, assisted in this unfathomably complex task by enormous computer models. (This, incidentally, is why some techno-futurists think that curing aging is better achieved by focusing on advances in computing power and artificial intelligence rather than biology. In reality, we will almost certainly need to do both—even the most advanced machine learning we could imagine running on inconceivably powerful computers needs real-world data on which to base its models.)

The logical endpoint of the process will be to gradually retire the idea of treating "aging" and begin to see all human dysfunction and disease as a "loss of homeostasis." Homeostasis is the collective term for the myriad processes which keep aspects of our physiology, from temperature and blood sugar to levels of proteins and numbers of a particular kind of cell, within the astonishingly narrow parameters needed to keep us alive. A 20- or 30-something human is in a state of very nearly perfect homeostasis, with odds of their system falling so far out of balance that they die less than one in 1,000 annually. If we could just return our physiological parameters to where they are in young adulthood, we'd be able to rely on our bodies' existing homeostatic systems to keep us alive.

The processes we currently label aging are a very gradual loss of homeostasis—far slower than, for example, the urgent need to start shivering when out in the cold to keep your body temperature within its safe range—but they reflect the fact that evolution doesn't need to maintain balance in our bodies into our sixth or seventh decade. The almost imperceptible unraveling of the near-equilibrium state we all enjoy in youth is why we become frail, forgetful and susceptible to disease. The best treatments for aging would gently nudge the network of processes causing us to gradually lose homeostasis back toward a stable state, keeping us safe and healthy for decades longer than we enjoy today. Intervening in clever ways to restore order to the whole system is surely the eventual future of medicine.

Unraveling the systems biology of aging is going to take incomprehensible quantities of data, enormous computing power and smart computational biologists working in tandem with those in the lab. Replacing narrative with numerical representation has revolutionized whole fields of science in the past, and the data and computational revolution in biology has only just begun.

Once we can model our biology in detail, we will be able to reprogram it to stop the gradual decrease in health and increase in risk of death with time. Human beings will finally be negligibly senescent, biologically immortal—ageless. The treatments that result will bring to an end the huge economic and human cost of natural selection's negligence, and the pain and suffering in old age which has been an inevitability for most living things for millions of years. It's a bold mission, but not an unachievable one: human biology is incredibly complex, but it is also finite. One day, data and powerful computer models will enable us to edit the very code we run on. Reprogramming aging will be our greatest achievement as a species. It should be our collective mission, as biologists, as doctors and as human beings.

PART III

Living Longer

9

The Quest for a Cure

Curing aging is a hugely important humanitarian goal which would alleviate suffering on the grandest possible scale. It will benefit every future generation of humans; if we don't blow ourselves up or upload ourselves into simulated brains in the next few centuries, it could benefit billions, even trillions, of people. There is no doubt that it is worth pursuing, especially now that we know it is scientifically possible.

But there's one question which everyone reading this book is at least wondering about, no matter how altruistic: will a cure for aging come in time for us? If not us, maybe for our children? And how can we move things along to bring forward the end of the age of aging? This final part of the book will explore that question. In the next few chapters, we'll take a look at what you can do now to give yourself the best chance of living a long and healthy life, and what scientists, doctors, governments and societies should be doing to maximize our collective chances.

The path to a cure won't be easy. We know from bitter experience that the journey from lab research to clinical medicine is tough, with many brilliant, elegant, robust ideas turning out to be a biological nightmare to put into practice. Turning findings from some long-lived mice into a drug or treatment is an endeavor that takes years, maybe decades, and millions or billions of pounds. It is often derailed by side effects, unforeseen differences between mice and humans, or just not working for no good reason. One potential benchmark is the billions and decades we've spent on cancer research without finding a cure. The goal sounds simple enough— "remove the diseased cells from the body"—but the challenges in implementation have been enormous.

However, there are lots of reasons to be optimistic. We've met many fundamentally different ways to extend the healthy lifespans of mice: dietary restriction, senolytics, telomerase, and even being sewn to a younger mouse. Though there's almost certainly a bit of mechanistic overlap given the interconnectedness of biology, this eclectic list should surely banish any remaining inklings that aging is inevitable, or that interventions which slow or reverse it are weird quirks of the lab environment. This expansive list of techniques suggests that aging is actually rather malleable, and there are many different ways to slow it down. Not only that, but it means we have lots of bites at the cherry—we'd have to be phenomenally unlucky for every variation of every single one of these diverse interventions not to pan out at all in people.

We know that all treatments have side effects, but some of the unintended consequences of the treatments we've discovered are positive. Senolytics, for example, improve conditions from osteoporosis to liver disease in mice, because the removal of senescent cells impacts so many different biological processes. Mitochondrially targeted antioxidant MitoQ, which we met in Chapter 7, improves the function of telomeres, but we also know that activating telomerase improves mitochondrial function. We saw in the last chapter that the genes **Klotho** and **FGF21** had an unexpected negative interaction, but some genes, rather than fighting, synergize. One particularly spectacular example was discovered in 2013 in **C. elegans**: if you take **daf-2,** whose mutants, you may remember from Chapter 3, live roughly twice as long as normal worms, and another gene called **rsks-1,** whose mutants live about 20 percent longer, and create worms in which both genes are mutated, they live almost five times longer than normal—an effect far larger than the simple sum of its parts. Because treating aspects of aging improves function in a variety of ways, we can hope for virtuous circles, where treatment **x** rejuvenates process **y** which in turn alleviates problem **z** which was never an intended target of **x** in the first place.

Finally, treating aging itself starts a virtuous circle of increasing life expectancy. At the moment, every year you're alive sees your probable date of death recede a few months into the future, and a significant part of this is because you're living through continuous advances in biomedical science. Those extra months carry on accruing, buying more time for more biomedical advances to add more months, and so on. We've already seen this effect work for previous generations: medicine, public health and hygiene meant that millions of children didn't die of infectious diseases in the 1930s, allowing the surviving 60-somethings to benefit from the new heart disease treatments of the 1990s and 2000s which weren't even on the drawing board when they were born. This cohort lived far longer than it would have done if medical progress had been frozen in the year of their birth.

If we can develop treatments against the aging process, people alive at the time the treatment is developed will get to live slightly longer, buying time to develop the next round of anti-aging medicine, and so on. As a bonus, it's likely that these treatments for aging will provide a greater uptick in life expectancy than a treatment for one specific disease will. We have already seen how even a complete cure for cancer would only add a few years to life expectancy, so a new, more effective treatment for a particular type of cancer will add even less because not everyone in the population gets that particular type of cancer. A treatment for aging, on the other hand, could delay all cancers, as well as heart disease, stroke and dementia, thus adding far more to life expectancy even if it's only partially effective.

The crucial moment comes if we can start developing and rolling out treatments for aging that mean life expectancy rises by one year per year. That would mean, on average, our date of death would be receding into the future as fast as we were all chasing it. If we can keep up this pace of innovation, that could continue to be true into the indefinite future—and it would be a de facto cure for aging. This idea is sometimes called "longevity escape velocity."

Whether a one-year-per-year increase in life expectancy is possible in a time frame which might be relevant to humans alive today is impossible to predict, and most scientists wouldn't blame you for being skeptical of its plausibility. However, the idea of developing treatments which buy us time to develop more and better treatments changes the nature of the challenge when it comes to adding substantial amounts to life expectancy, or even, one day, curing aging entirely.

A "cure" for aging as we usually imagine the word would be a single treatment which totally stopped your body from aging at all. That is, quite simply, impossible given our present state of knowledge. It would need us to have the full systems biology understanding of aging we discussed in the previous chapter, today—to reengineer human biology from the ground up. If instead we can "merely" increase life expectancy through conquering aging piece by piece, we can develop a cure in practice, but without needing to be nearly so clever to do so: we only need to be smart enough to stay one step ahead of aging. Even if you remain skeptical of anyone touting a time frame, it's undeniable that this approach vastly reduces the complexity of the problem and brings forward the date of this medical revolution dramatically.

However and whenever we do manage to cure aging, this is how it will happen: the cure for aging will be a jigsaw of treatments which evolves with time, a succession of technologies which gradually improve life expectancy to the point that people will notice that they've stopped aging—not a miraculous magic bullet discovered in a flash of insight by a lone genius. The first ageless generation probably won't realize their luck at first—they'll grow up expecting to die at 100, or 150, or whatever "old" is for their society but, one after another, lifesaving medical breakthroughs will push their funerals further and further into the future. It will be very hard to call when we've cured aging if you live through it, constantly wondering if the next breakthrough could be the last and life expectancy will finally stall—but it will be blindingly obvious with hindsight, surveying centuries of life expectancy statistics, and spotting the point where people just stopped dying of old age.

So, if you want to live a very long time, what you have to hope is that our first generation of anti-aging therapies can tide us over for a few decades until systems medicine can begin to suggest more nuanced treatments, which then buy us another few decades for those treatments to improve. It's far from an absurd hope: senolytics could be just a few years away, albeit most likely for specific diseases rather than the aging process itself at first. More advanced treatments like gene and stem cell therapies could be available on timescales measured in decades— soon enough to matter for many of us. One way or another, we will eventually start adding a year or more to life expectancy every year—the only question is when, and how we can maximize our chances of still being alive when it does.

That's what the next two chapters will cover: first, how to maximize your personal chance of living as long as possible, looking at how the science of aging translates into health advice; then, what governments and society need to do to make this biomedical revolution happen as soon as possible. All of these tips are win-win: the worst-case scenario is that you live a longer, healthier life sufficient to benefit from as-yet-uninvented medical treatments, and bring forward the date of aging being cured for our children, or their children; the best case is that some people alive today might live dramatically longer in good health than we currently expect. Let's look at how to make that happen.

10

How to Live Long Enough to Live Even Longer

With only a small fraction of life expectancy explained by genes, most of your longevity is down to lifestyle and luck. Luck is, by definition, impossible to do anything about—but there are plenty of scientifically backed suggestions to max out your life expectancy based on how you choose to live.

The potential dividends of optimizing your lifestyle are large. One study looking at 100,000 health professionals in the U.S. gave them a score based on five healthy behaviors (not smoking, a healthy bodyweight, not drinking too much, regular exercise and eating well), and found that those who ticked four or five boxes at age 50 could expect to live ten years longer, both in total and in years spent in good health, than those who didn't tick any. Around 40 percent of cancer and a staggering 80 percent of cardiovascular disease is thought to be preventable, meaning that, if we all lived an optimal lifestyle, cancer and heart problems would be dramatically deferred. Deciding whether to take a new anti-aging pill is a hard decision which will involve weighing up complex evidence as it comes in—trying to live a bit more healthily is a no-brainer.

It's important to remember how hard it is to pin a disease of aging on any single cause. Imagine you acquire ten mutations which cause a particular cancer, and three were caused by alcohol, one by food and six by unavoidable random DNA damage from being alive. Should you chalk it up to bad lifestyle, or bad luck? It's like **Murder on the Orient Express**: no individual actor is really responsible. Living well improves your odds, but it can never reduce them to zero. The more optimistic perspective is that it should absolve you of overly detailed soul-searching for the precise cause if you do get diagnosed with cancer or have a heart attack, while at the same time empowering us all because lifestyle improvements really can help—even cancer, which can seem completely out of our hands, is not entirely a matter of chance.

It's also important to mention that it's never too late to start with any of these health tips, because the changes in our bodies as we age are cumulative. If a change to your lifestyle slows the accumulation of mutations in your DNA that make cancer more likely, it will help no matter how old you are when you start: if a cell teetering on the brink of cancer just needs one more mutation, even preventing that single extra change could save your life. Studies on exercise programs bear this out—exercising improves health even in 80-somethings, and it's often those who are the least healthy to begin with who benefit the most. The best time to plant a tree was 20 years ago, as the saying goes—but the second-best time is today.

The advice which will help you live the longest is surprisingly basic in some ways, but it's also worth saying that following this advice isn't always easy, and requires both opportunity and willpower. However, now that you know a lot more about the biology of aging than before, I hope that understanding the science behind them will make even familiar suggestions far more compelling.

1. DON'T SMOKE

Smoking is outrageously bad for you. If you want to live a long life in good health but you smoke, the first thing you should do is quit.

Those who smoke throughout their lives have their life expectancy slashed by around ten years. Smokers can't even claim to live fast and die young: they experience about the same number of years in ill health at the end of life as non-smokers, meaning a greater fraction of their shorter lives over-all. Smoking is responsible for 90 percent of lung cancers and almost half of deaths from lung disease. And, while the lungs bear the brunt of the assault from cigarettes, there's a reasonable case to be made that smoking basically accelerates the whole aging process: it also increases the risk of many other cancers, plus other diseases of aging like heart disease, stroke and dementia. It even makes you **look** older, causing thinning skin, wrinkles, hair graying and baldness.

Cigarette smoke contains hundreds of toxic chemicals which cause mutations in your DNA. They leave a specific "mutational signature" in the DNA of cancers they cause—the lining of smokers' lungs contains a lot of mutations where a C has transformed into an A, for example. This and other mutational signatures are found in tissues around smokers' bodies because the chemicals are absorbed into your blood, allowing them to affect far more than just the lungs. These extra mutations give cancer many more rolls of the dice, causing the sequence of clonal expansions we discussed in Chapter 7 to proceed at a faster rate, and increasing the risk of the disease.

Smoking causes chronic inflammation, which is thought to be behind the smoking-related increase in cardiovascular disease. Remember that atherosclerotic plaques are mainly composed of dying immune cells—agitating the immune system accelerates their formation. Smoking also causes cellular senescence, shortens your telomeres and even increases the formation of AGEs (the advanced glycation end products that form when sugars react with proteins) in our tissues, in part because of highly reactive chemicals in cigarette smoke.

The good news is that quitting can quickly reduce your risk, or even return it to normal. Inflammation falls rapidly after giving up, and reaches normal levels around five years after stopping smoking, along with the risk of cardiovascular disease. Overall, quitting can add years to your life: even stopping smoking at 60 will increase your life expectancy by about three years, while quitting at 30 pretty much restores your life expectancy to normal.

2. DON'T EAT TOO MUCH

It won't come as a surprise that what you eat can have a significant effect on how long you live. Getting a balanced diet with plenty of fruit, vegetables, whole grains and nuts is important and can add substantially to your healthspan and lifespan. Exactly what combination of foods is optimal is extremely hard to establish and, as a result, hotly contested. The ideal experiment, in which you randomized thousands of people to different relative quantities of a variety of foods for multiple decades, would be impractical, hugely expensive and probably unethical. This means that scientists are left to pick through observational studies—and, since people's eating habits are tied up with their wealth, social status, general interest in health and genetics, all of which will also influence their lifespan, it's very hard to pick apart cause and effect.

The result is that the best advice is probably pragmatic: eat a good mix of different foods, not too much of any one thing, while watching the amount of very sugary, fatty or processed food you eat, and not drinking too much alcohol. Binging on the latest "superfood" almost certainly won't transform your health, but maintaining a healthy balanced diet certainly can.

It might also be beneficial to cut down on your meat consumption if you eat a lot. The observational evidence in favor of vegetarianism is suggestive but not entirely clear, but there are several biological mechanisms by which a plant-based diet could be better for you. Eating more fruit and vegetables has been shown to improve the diversity of your microbiome. Obtaining your protein from plant rather than animal sources may also cause a kind of dietary restriction. Plant proteins have a different ratio of amino acids—the building blocks of proteins—which is less optimal for human needs than animal protein, but, ironically, this could be better for us thanks to amino acid restriction being a type of DR. Finally, tiny quantities of defensive toxic chemicals produced by plants might actually be slightly poisonous to us—but to a sufficiently small extent that our bodies overcompensate when removing them and repairing the damage they cause, making us healthier overall. The concept that a little bit of stress can induce stress responses which actually make us healthier is called "hormesis," and might be applicable to other lifestyle tips too.

However, there is one key finding when it comes to diet and lifespan: a large body of evidence suggests that excess fat is bad for you. Some research has found that being a little overweight is good for long-term health, but they, too, are observational studies mired with confounding and complicating factors: people who are underweight, especially in old age, have often lost weight due to illness; weight is related in complex ways to socio-economic status, which has profound effects on health; body mass index (aka BMI), which is usually used in these studies, is too simplistic a measure; and so on.

Overall, however, it seems likely that shedding a few pounds would lower most people's risk of many of the diseases of old age, all at once. In spite of its limitations, looking at the life expectancy of people with different BMIs does give us some idea of the opportunity here. BMI is calculated by taking your weight in kilograms and dividing it by your height in meters, squared. The "normal range" is often quoted as 18.5–25 kg/m². (If you prefer to use pounds and inches, then the formula is weight in pounds divided by height in inches squared, times 700.) Having a BMI over 25 kg/m² is "overweight," and might knock a couple of years off your life expectancy; more than 30 kg/m² is "obese" and will lower your lifespan by a few more; and being heavier still can knock a full decade off your life. Being overweight will probably knock even more years off your healthspan: the increased risk of heart disease and diabetes means heavier people are likely to spend their later years in poor health. In fact, obese people not only tend to live less long, but actually incur greater health expenses in spite of their shorter lifespan. All this suggests that carrying excess weight is worth avoiding.

Years spent overweight conspire to basically accelerate aging. This is because fat isn't the passive store of energy we might imagine it to be—fat deposits are more correctly known as adipose tissue, which is made up of cells called adipocytes whose job it is to store the fat. Where the adipose tissue is located makes a big difference to the effects it has. "Subcutaneous" fat lies just below the skin (the stuff you can grab), and "visceral" fat accumulates deep inside our bodies in the spaces between our organs (which I would definitely not recommend grabbing). Visceral fat seems to be by far the worse of the two, emitting pro-inflammatory molecules which fuel chronic inflammation. It might not actually be the adipocytes themselves that are responsible for the inflammatory overload, but immune cells which reside between them—but this distinction doesn't really matter to our aging bodies.

The importance of visceral fat in this process is the reason why it's worse to be "apple-shaped"—to have a fat belly—than "pear-shaped"—carrying fat on your hips and buttocks. The rounder, "beer belly" look results from fat in our abdomen between our organs, inflating our tummies from deep within— in other words, inflammatory visceral fat. Fat on the legs, hips and bottom, by contrast, is mainly stored safely just beneath the skin as the relatively benign, subcutaneous stuff. Before the menopause, women tend to accumulate fat more on their buttocks and men more on their tummy, though obviously it varies. You can, of course, have both.

Simple statistics based on body measurements provide circumstantial (and circumferential) evidence against visceral fat. In the epidemiology literature, BMI's shortcomings mean that it's currently battling a number of other measures and ratios as best predictor of age-related ill health. A top contender is waist-to-height ratio which involves measuring around your waist and dividing it by your height, in whatever units you like as long as they're both the same. Its normal range runs from 0.4 to 0.5, though some people do score slightly lower. Also, if you're over 50, the normal range is relaxed slightly and it's probably okay to edge up a little to 0.6 without adding substantially to your risk of getting ill.

The most well-known objection to BMI is that muscle is denser than fat, meaning that if you're especially muscular you can end up with a BMI which tips you into the overweight range without actually being fat. (Unfortunately for most of us, you have to be pretty ripped to use this as an excuse.) Another trick BMI misses is distinguishing subcutaneous and visceral fat—BMI doesn't care if you're carrying excess kilos under your skin or around your organs, reducing the accuracy of its health predictions. Waist-to-height ratio improves on this—your waist circumference is related to how much padding there is in your interior. Accordingly, some studies show that it does better at predicting the likelihood of having a heart attack or developing diabetes than BMI.

Being overweight also puts you at risk of diabetes, which exacerbates many of the problems of aging. If you are diabetic, evidence suggests that losing weight and reducing your blood sugar levels may be able to get these problems under control, dramatically reducing the risk of a whole swathe of illnesses that diabetes exacerbates if you can manage to keep the weight off.

There's also evidence that cutting back on sugar can have health benefits, in part because it will reduce the quantity of glycated proteins in your body. It might be that some sugars are worse than others in this regard—fructose reacts more readily with proteins in a test tube, for example, and it's been suggested that it could be a more potent initiator of glycation when eaten, too. Another, more speculative suggestion is that it might be worth cutting back on AGEs—the advanced glycation end products that sugars form when they react with proteins—in your diet. We still haven't nailed down the chemistry behind the formation of AGEs on our proteins, and it's possible that any AGEs you consume ready-made may be able to stick to your collagen and other proteins and cause problems. This means avoiding food cooked at high temperatures which accelerate AGE-ing, such as fried or roasted foods, and preferring to eat things that are raw, boiled or stewed. However, given their ubiquity and deliciousness, dietary AGEs are quite hard to avoid.

If you've already dieted and exercised your way to a healthy weight, you might be wondering if you should go further. I promised in Chapter 3 that we'd return to the topic of dietary restriction in humans. We should start by examining the results of DR in the closest evolutionary relatives it's been tried in: rhesus macaques. In 1987 and 1989, experiments were launched by the National Institute on Aging (NIA) and the University of Wisconsin–Madison with a total of just under 200 monkeys.

First, the good news for DR: monkeys fed a normal diet lived 21 years without disease, while the DR groups enjoyed an extra five-plus years disease-free in both studies. If this result translated into humans, with the simplistic assumption that every monkey year is two or three human years, that could easily mean an extra decade free from age-related disease. However, the results for longevity are rather more confusing. The Wisconsin macaques showed a clear benefit of DR: the control group lived an average of just over 25 years, while the calorie-counters made it to nearly 29. Again, translated naïvely into human years, that's not far off a decade. By contrast, the NIA monkeys, whether DR or control animals, had statistically indistinguishable lifespans.

Perhaps the simplest explanation for these disappointing results is that there are diminishing returns to DR in longer-lived animals like monkeys (and people). Once you've got a basically healthy diet, the theory goes, there's less to be gained by restricting your food further. The Wisconsin monkeys' diet was made up of pellets of protein, sugar, oil and vitamins, relatively high in fat and sugar, and the non-DR monkeys were allowed to eat as much of it as they liked—a problem perhaps compounded by the pellets being so deliciously fatty and sugary—making the contrast between them and their DR compatriots stark. By contrast, the NIA monkeys were fed a mix of beans, grain and fish which was higher in fiber, lower in fat and sugar, resulting in a healthier diet for the control animals.

Thus, the monkeys' collective diets lie on a sliding scale of quality, running from the unhealthy Wisconsin control monkeys, via the Wisconsin DR and NIA control groups whose diets were in the middle somewhere, to the NIA DR monkeys, whose food was both the healthiest and the most restricted. By this logic, there's an advantage to moving from dietary excess—the monkey equivalent of all-you-can-eat hamburgers and sugary soft drinks—to a more modest intake; but eating less still, like the NIA DR group did, won't add much to your lifespan if you're already eating a moderate amount of nutritious food. That said, the NIA's monkeys did still gain five extra years disease-free—something most of us would jump at, even if we didn't live longer as a result.

Hawkish proponents of DR contend that these studies were flawed and understate its potential benefits, and technical differences between the two experiments provide enough fuel for years of debate. However, the fact that the results are hard to interpret is itself evidence that we can't hope for a massive effect in animals like us—if DR nearly doubled lifespan like it does in nematode worms or rats, a lack of perfect equivalence between two experiments shouldn't mask such a huge effect.

There have been studies in humans: DR results in significant weight loss and improves markers of health (things like blood pressure, cholesterol, levels of inflammation, and so on), but we don't know how it affects lifespan because the studies so far have been too short to check. The other line of (non-)evidence for DR in humans is observational. Throughout history, individuals, societies and religions have practiced many and varied diets, out of choice and necessity. One oft-cited example is the unusual longevity of people in Okinawa, a tropical island southwest of mainland Japan, which has been pinned on their culture of nutrient-dense but low-calorie diets. However, it's hard to be sure—there may be other cultural or genetic effects at play which are unique to the island. And the effect isn't that dramatic anyway: Okinawans only live about a year longer than people in the rest of Japan, rather than reliably attaining triple-digit lifespans. The effect is also slowly disappearing, which is thought to be due to the increasing westernization of the islanders' diet.

DR also comes with side effects. DR mice with flu die more often than their peers on normal diets—eating less, even with optimal nutrition, seems to be bad for the immune system. In humans, a few participants in one trial were forced to stop due to anemia (a condition where a lack of red blood cells or oxygen-carrying hemoglobin leaves your organs short on oxygen) or significant drops in bone density. DR practitioners also report feeling the cold more easily, increased irritability and lower sex drive. DR mice might live longer in a clean, calm lab environment, kept meticulously free from infection—but as a human who wants to get out and about and actually do stuff, it's no use being at slightly lower risk of age-related disease if you end up frequently breaking your leg, or dying young of influenza.

When you eat as well as what you eat can have an effect. "Intermittent fasting," recently popularized as the "5:2 diet," involves eating far less, or perhaps nothing at all, for a day in every few. The 5:2 diet, for example, suggests cutting back to 600 calories on two non-consecutive days in a week, and eating normally on the other five. "Alternate-day fasting" goes a little further and requires low or zero eating every other day. "Periodic fasting" means not eating for five-plus days in a row, anywhere from once a month to annually. Finally, "time-restricted feeding" confines eating to a window typically ranging from 6 to 12 hours a day.

Evidence on these different DR variants is harder to unpack even than for DR overall, due to less experimental data. The practical idea behind them is that they're easier to stick to, allowing devotees to eat as normal when not fasting, and only endure limited rather than continuous bouts of hunger. More theoretically, arguments rage about whether fasting works by the same mechanisms as "conventional" DR or subtly different ones, which governs whether we'd expect the effects to be equivalent or not.

So, should you take up one of these draconian diets? The short and infuriating answer, in spite of decades of experiments, is that we simply don't know. It feels like we're tantalizingly close to an answer—the more you read about DR, the harder it gets to shake the nagging feeling that there is some optimal diet out there, waiting to be found. The enduring popularity of fad diets and "superfoods" reflects our intimate, ambivalent relationship with eating, and a deeply held intuition that what we put into our bodies can have a profound effect on health and lifespan. Food is an emotive, engaging topic for many of us. From a more practical standpoint, DR and fasting are also very simple and cheap—you could literally start doing either right now and, far from being costly to implement, it would probably save you money because you could buy less stuff during the weekly shop.

However, I find it hard to recommend a diet which might leave you immune-compromised, thin-boned, chilly and irritable. Certainly if you were thinking of trying DR or any of its variants, you'd be well advised to talk to your doctor first, and set up some monitoring for the potential side effects. I don't think we'll ever know the full story about DR, or optimizing our diet in general, before it's too late: there are so many variables in what and when we eat that I think we'll cure aging and render the fine details of diet irrelevant before we fully figure it out.

It's important not to obsess over optimizing diet, chasing some elusive, possibly nonexistent perfection, and lose sight of the big picture. Though the evidence is complex, we can't throw our hands in the air and gorge on all the cake we like: unless you're already skinnier than most, losing weight will probably do you some good. Both as individuals choosing what we eat and as a society choosing policies to adopt, we should be looking to reduce obesity. Helping overweight people into the normal range of BMI is both easier and more unambiguously beneficial than trying to get already healthy people to engage in DR.

So, try to eat a balanced diet, and not too much of it. Though the diet debate will no doubt roll on indefinitely, the key findings are that variety is good, and that being overweight is bad for you because it literally accelerates the aging process.

3. GET SOME EXERCISE

Exercise is good for your health, and it doesn't need to be in intimidating amounts: studies show that every minute's increase in exercise or decrease in time spent inactive per day reduces your risk of death. Exercise is also known to reduce the risk of dozens of diseases, including the most significant age-related ones which you can no doubt reel off by now—it even delays cognitive decline and dementia. It also seems likely that the first, small steps are the most significant, with increasing time or intensity providing diminishing returns. Starting small by trying to fit a five- or ten-minute walk in every day pays dividends for your health and helps to make doing a bit more seem less intimidating over time.

If your lifestyle is entirely sedentary, even half an hour per day of light physical activity can reduce your odds of death by 14 percent. Ten to fifteen minutes of moderate exercise a day goes further, roughly halving your risk of death from any cause. Exercising for 30 minutes a day helps a bit more. Benefits beyond this are unclear—most studies find something of a plateau and perhaps even a small **increase** in risk, but it's hard to be sure because so few people do sufficiently huge quantities of exercise that it's difficult to draw statistically robust conclusions. In any case, there's no quantity of exercise which is riskier than doing none at all, meaning it's not an excuse to stay sedentary—but, if you're already running for an hour every day, it's unlikely you'll experience any benefit from upping that to 90 minutes.

Studies do find that Olympic athletes have lower mortality than the general population, but it's not clear that this is down to how much exercise they do. First, it might be that the causation runs the other way here: perhaps Olympians are able to endure punishing training regimens and compete at such a high level **because** their bodies are more robust than the average person's. Second, there may be entirely different mechanisms at play: chess champions also live longer, by a similar amount to elite athletes; and Nobel Prize winners tend to outlive similarly high-flying scientists who were nominated but didn't win the award by a year or two as well. These studies are intriguing, and suggest that acclaim itself may be medicinal—undermining the simple idea that extreme physical fitness alone is behind Olympians' longevity.

For us mere mortals, it's not just cardio which pays dividends—resistance training can help fend off the decline in muscle size and strength experienced by older adults too. We lose about 5 percent of our muscle mass and 10 percent of our strength every decade after the age of 30, and the rate more than doubles after age 70. This decline is called sarcopenia—the technical term for age-related loss of muscle mass. Research shows that this loss of strength can be substantially reversed by resistance training. As usual, it's never too late to start: exercise programs have been shown to improve the health of 90-somethings, with a two-month program of resistance training nearly doubling nonagenarians' muscle strength, and increasing their walking speed by 50 percent.

When you exercise, there are dozens of changes to your metabolism, circulation, bones, and even the nerves which connect your brain to your muscles. Exercise increases the length of telomeres, reduces the number of senescent cells in muscle and increases the number of "satellite cells" (the stem cells that renew muscle), and also increases the activity of other stem cells around the body. Exercised muscle contains more and better mitochondria than sedentary muscle does. Exercise may result in the destruction and rebuilding of aging collagen, meaning that stiff, glycated fibers are replaced with fresh new ones. It also reduces inflammation: taken together, our muscles are the biggest organ in our body, meaning that the signals they secrete are significant in volume. Static muscle tends to promote inflammation, while active muscle does the opposite. Getting fit also has an indirect anti-inflammatory effect by burning fat which, as we've just learned, also secretes inflammatory molecules.

The benefits of exercise are so wide-ranging and systemic that doctors quip that if exercise were a drug, everyone would be queuing up to take it. Unfortunately, exercise is harder work than popping a pill, and our hectic lifestyles don't always make it easy—but it really is worth taking the first step, even if you start small.

4. GET SEVEN TO EIGHT HOURS OF SLEEP A NIGHT

Getting seven or eight hours of good-quality sleep per night is probably optimal for health, but it's hard to be totally sure because sleep is a very tricky thing to study. Large systematic reviews find that getting less than this is associated with an increased chance of death and—less widely publicized—that getting more than eight hours of sleep is associated with a **larger** increased risk of death than getting too little.

The challenge is working out whether this relatively robust finding suggests causality: are people sleeping 11 hours a night doing so out of choice, or is it more likely that they've got an underlying health problem which means they need the extra sleep? Is the short lifespan of people sleeping four hours a night because of a lack of rest, or because they live stressful lives which impact their health and also coincidentally reduce the time they have to sleep?

Short of enforcing sleep durations on people for decades in the name of science, the best answer to this question would be to identify biological mechanisms that could connect sleep and longevity. We met one suggestion for which evidence is growing in Chapter 5: while we sleep, our brains take the opportunity to spring-clean, including flushing out the toxic amyloid which is implicated with Alzheimer's disease. That's a good incentive to try to stop a box-set binge one episode early and get a bit of extra shut-eye.

Sleep might also be one place where feedback processes can worsen the problems of aging. Older people tend to sleep less well, while the healthiest older people have a stable sleep rhythm. If aging worsens your sleep, which in turn worsens your health, this could make for a vicious cycle. One example is the clouding and discoloration of the lens of the eye resulting from protein modifications, the process behind the formation of cataracts. These tend to absorb blue light, making the world around us seem to take on warmer hues. We now know that our eyes use light levels, in particular levels of blue light, as a driver of our circadian rhythms. This is the logic behind "night mode" on computers and phones which makes the screen dimmer and more orange in color—bright blue light during the day tells our brain that it's time to stay awake, and reducing exposure to blue in the evenings can therefore be beneficial for our sleeping patterns. Older people, with lenses tinted orange by protein degradation, are naturally exposed to less blue light regardless of time of day, undermining this subtle physiological cue. Sleep quality tends to improve after cataract surgery—removing the yellowed, cloudy lenses can restore not just our vision, but also our circadian clock's blue cues.

Though the evidence isn't yet totally watertight, getting a good night's sleep may well improve your healthy lifespan—and make mornings more tolerable as a pleasant side effect.

5. GET VACCINATED AND WASH YOUR HANDS

Vaccinations are one of the most important ways that humans have reduced mortality throughout our lives, and getting them doesn't protect just you, but also those around you. Not only do they mean you're likely to live longer because you won't die of whatever infectious disease they prevent, they will also reduce your lifelong burden of inflammation, which can slow aging, too.

If you've had all your childhood shots, the most

common vaccine that adults need is against seasonal flu. Many countries have an annual "flu season," usually lasting a couple of months in the winter, though its length and severity can vary quite significantly from year to year as different strains of flu ebb and flow.

If you're an older adult, getting a flu shot is well worth it. People aged 65 or over are ten times more likely to be hospitalized with flu, and around 20 times more likely to die of it than those aged 18 to 64. Just counting direct deaths from flu probably underestimates its true impact, especially in older people—deaths from heart attacks, strokes and diabetes also peak around flu season, and there's evidence that flu is the trigger. Even though the vaccine gets less effective with age, the risks of flu are so dire that it's worth getting. Giving precise numbers is difficult because flu vaccines are considered sufficiently effective that it would be unethical to purposely not vaccinate a subset of your older patients in studies.

Even in younger adults, the calculus is fairly clear because the flu shot is cheap, and has a decent chance of stopping you from needing to spend a week wiped out in bed with fever, muscle ache and total exhaustion. The vaccine's side effects are also fairly innocuous: it can cause mild flu-like symptoms, or a bit of aching at the site of injection. Throw in reduced inflammation and protection of any older relatives or people in your life who can't get the vaccine, and there's a pretty compelling case to get vaccinated however old you are.

It's also worth following standard advice to avoid infections: wash your hands thoroughly and regularly, cook food thoroughly and take time off work if you're unwell—this won't just improve your colleagues' healthy lifespans, but could have a much wider impact if it stops them from passing the disease on to others, and so on. Of course, there could be no better example of the importance of basic hygiene and nipping transmission chains in the bud than the coronavirus pandemic.

It might even be worth avoiding infections to optimize your aging more generally. There's evidence that historical progress fighting infectious diseases in youth had additional, indirect effects on life expectancy, with children who faced fewer infections growing up being at less risk of diseases like cancer and heart disease in old age. The hypothesis is that the reduced burden of infection reduces the cumulative burden of inflammation, slowing the aging process throughout life.

Infections can also directly cause seemingly unrelated diseases. In some cases the link is very clear, such as human papillomavirus (or HPV) causing cervical, mouth and throat cancers: HPV is now widely vaccinated against, primarily to reduce the risk of cancer rather than because the infection itself is particularly unpleasant. Another example is **Helicobacter pylori,** the bacterium which causes stomach ulcers and a substantial fraction of stomach cancers. There are also suggestive reports of bacteria and viruses being found in the plaques which clog our arteries in old age, and the brains of patients with dementia. Whether these bugs are the cause of these conditions, aggravate them, or are merely opportunistic or even innocent bystanders remains to be fully elucidated.

Overall, it's probably worth taking reasonable measures to steer clear of infectious disease for benefits beyond avoiding the immediate misery of being ill.

6. TAKE CARE OF YOUR TEETH

You've probably been told one hundred times by your dentist: brush twice a day with fluoride toothpaste, clear out the gaps between your teeth with floss or interdental brushes and avoid sugary snacks and soft drinks. What you might not know is that your dentist's advice will affect more than just your smile and future dental bills—it can impact your lifespan, and even your risk of dementia.

This first came to light in a series of studies in the 1980s and 1990s which sound like examples you'd use to illustrate the problems with observational studies. Epidemiologists noticed that people with tooth decay and gum disease were more likely to develop heart disease as they got older. This sounds like a classic case where correlation doesn't imply causation: perhaps some people have less time and money to spend taking care of their diet, getting enough exercise and looking after their teeth; or maybe people who are less health-conscious generally both eat unhealthy food and don't bother brushing their teeth afterward. These explanations suggest that bad oral hygiene and heart problems would appear together, but with neither causing the other, both being caused instead by a third, unmeasured variable, like poverty.

However, this relationship seemed to hold up even as the statisticians tried to correct for these confounding factors. One study found that people who brushed their teeth twice a day were at lower risk of heart attack than those who brushed them once, who were in turn safer than those who didn't regularly brush. They also showed a similar relationship for levels of C-reactive protein (the blood test for inflammation which is usually slightly raised in older people)—the more frequent brushers had less in their blood. This suggests a "dose–response" relationship, where doing more of something (often, taking more of a drug in a clinical trial) has a larger effect—and, while it doesn't prove that poor dental hygiene **causes** heart attacks, it does make it more plausible. It's also been shown that the types of bacteria present in your mouth can have an effect on both diabetes risk and life expectancy, even if they don't give rise to levels of gum disease which would alarm your dentist.

The proposed biological link here is chronic inflammation. The ongoing battle with bacteria in your mouth that cause chronic gum disease, tooth decay and so on, even at low levels, results in a constant fizzing of inflammatory molecules. This, as is now our common refrain, basically accelerates the aging process. In Chapter 5, we even met reported links between gum problems and Alzheimer's disease, with bacteria responsible for gum problems being found in amyloid plaques. While these theories aren't proven yet, it's another good reason to keep your teeth clean.

7. WEAR SUNSCREEN

We mentioned in Chapter 4 that skin aging is very closely related to its exposure to the sun. Sun-exposed skin wrinkles more rapidly, is at risk of developing the mottling and discoloration we associate with aging and, less cosmetically, is at significantly higher risk of transforming into a skin cancer. Getting sunburn just once every two years is associated with increased cancer risk.

All of these phenomena are the result of ultraviolet light present in sunlight. UV light has enough energy to break apart the chemical bonds holding molecules together—including proteins and DNA. Damage to DNA, if repaired incorrectly, can turn into a mutation and risks putting a cell on the path to becoming cancerous. Damage to proteins like collagen and elastin, which make our skin supple, can make it stiffer with age.

As a result, blocking UV from reaching your skin can stop the aging effects of sunlight. You can do this by not going outside when the sun is particularly high in the sky, covering up exposed areas with clothing, or applying sunscreen which absorbs the ultraviolet light. Among a cornucopia of skin creams which claim to be "anti-aging," sunscreen has by far the best scientific evidence behind it.

8. MONITOR YOUR HEART RATE AND BLOOD PRESSURE

There's an increasing proliferation of apps and devices to quantify every aspect of your life, but probably the most valuable is the humble automated blood pressure cuff. By measuring your heart rate and blood pressure you can get a significant insight into the state of your cardiovascular health—which, given that heart disease, stroke and vascular dementia are common causes of death and disability, is a significant insight into your health overall.

Every heartbeat ejects a surge of blood into your aorta—your central artery. The circulatory system is structured like a tree: the aorta is the trunk, while increasingly smaller vessels are the branches and twigs, delivering blood to every corner of your body. A blood pressure monitor gives you two numbers, like 120 over 80 (both are measured in a slightly archaic unit of pressure, millimeters of mercury). The first, larger number is called the systolic pressure, and it measures the pressure wave which spreads throughout your body from your heart as it beats; the second, smaller number is the diastolic pressure, which is the minimum pressure in your blood vessels between heartbeats. Arteries with soft, elastic walls can absorb the force of the heart's pressure wave, meaning that the ever-smaller vessels further from the heart experience less of it. Glycation and loss of collagen and elastin, atherosclerotic plaques, TTR amyloid and other processes make blood vessels narrower and stiffer—and these inflexible arteries transmit the shockwave's full force. Those same processes also make the vessels more brittle, and the final, tiny vessels are extremely delicate: pummeling them with too high a pressure over and over, 60 to 100 times a minute, all day, every day, can eventually cause them to burst.

The most serious and sudden side effects of a burst blood vessel occur if it affects a medium-sized vessel in the brain, causing a bleeding, or hemorrhagic, stroke. This leads to blood pooling rather than flowing through that part of the brain and, within minutes, nearby brain cells start to die from lack of oxygen. It's also possible for smaller vessels to burst, which may not be immediately noticeable, but many small events over time can contribute to vascular dementia. High blood pressure also damages delicate structures in the kidneys that filter our blood, can cause blood vessels in the back of the eye to dilate or burst, and also has more unexpected effects like reducing bone strength.

High blood pressure, also called hypertension, is a silent killer. Globally, around 40 percent of people over the age of 25 suffer from it—but you can't feel high blood pressure, nor does it have any immediate symptoms. That's why you need a blood pressure cuff: sit down, relax, take a few deep breaths, take a measurement, and keep a note so you can watch for trends over time. Blood pressure is considered normal if it's below 120/80. From about 115/75, every additional 20/10 roughly doubles the risk of death from heart disease or stroke—so 135/85 means twice the risk, 155/95 is four times as dangerous, and so on. If your blood pressure is regularly above 120/80, it's probably worth trying to improve your diet or exercise a bit more—these simple interventions are an excellent way to reduce blood pressure. If you consistently get results of 140/90 or more and your doctor doesn't already know about it, it's worth scheduling an appointment to discuss it with them, and considering starting on medication. Home blood pressure readings are invaluable because a lot of people get markedly higher readings when they're done by a doctor—a phenomenon delightfully known as "white-coat hypertension."

It's also worth keeping an eye on your heart rate. Most automated blood pressure cuffs will give you a heart rate readout at the same time as they measure your blood pressure. Your "resting heart rate" should be somewhere between 60 and 100 beats per minute, though if you're very fit it can drop a bit lower. As we noted in Chapter 4, a resting heart rate of 100 bpm instead of 60 approximately doubles your risk of death. Intriguingly, the risk that doubles is death from any cause, not just heart disease—a high resting heart rate also correlates with increased risk of cancer. The prescription is similar to that for high blood pressure: losing some weight and getting more exercise can bring a fast-beating heart down to a healthier rate.

9. DON'T BOTHER WITH SUPPLEMENTS

Unless you have a specific vitamin deficiency which you need supplements to treat, the evidence doesn't support the use of the various vitamin pills on the market. As we noted in Chapter 7, a gold-standard review of trials involving almost 300,000 people found that vitamin supplements either had no effect on risk of death or, in the case of beta-carotene and vitamin E, slightly increased it.

In spite of this damning synthesis and the decades of failed trials it summarizes, the antioxidant myth persists: supplements remain popular with the public, with around half of U.S. adults reporting that they take them regularly. Vitamins sound healthy, and popping a pill is easier than improving your diet or doing more regular exercise; but you'd be far better off spending any money you would have used for supplements on vegetables, or saving up for a pair of running shoes.

10. DON'T BOTHER WITH
LONGEVITY DRUGS—YET

If you have a health condition, medication can literally keep you alive—and it's obviously best to weigh the costs and benefits of a particular treatment with your doctor—but, for people who are in generally good health for their age, there's not yet a pill that can extend your healthy lifespan.

It's sometimes suggested that we should all be taking a daily baby aspirin, which could theoretically be somewhat anti-aging by reducing inflammation, and with it the odds of heart attack or stroke. Unfortunately, any dose of aspirin comes with a risk of stomach bleeding, meaning that you have to be at increased risk of heart attack before daily aspirin is worth taking on balance—and, even then, opinions differ within the medical profession as to whether the costs outweigh the benefits.

The diabetes drug metformin is a top contender to slow the aging process; it's one of the DR mimetics we discussed in Chapter 5. Though the evidence so far is promising, a large trial in the U.S. will give us a clear answer one way or the other in the next five years or so—it's probably best to just wait it out. (We'll talk more about this trial in the next chapter.)

Another idea to watch is radical reduction of cholesterol levels, most likely with PCSK9 inhibitors, as we discussed in the previous chapter. Evidence so far suggests that humans can get away with far lower levels of cholesterol in our blood than is currently the norm and, if PCSK9 inhibition proves safe in the long run, these drugs (or the follow-up "cholesterol vaccine" gene therapies) might be worth taking. However, more work is needed to establish that this is truly safe, especially in people without elevated cholesterol levels.

As some of the ideas we've discussed in the preceding chapters become reality, many people of different ages will be watching keenly, trying to work out if the current weight of evidence for a treatment means it's worth taking. Coming up with mechanisms to allow us to make informed decisions about this is vital—and we'll be exploring that in the next chapter, too.

11. BE A WOMAN

We'll finish with arguably the least helpful piece of advice: being a born woman can improve your life expectancy by somewhere around five years. There are a variety of social factors which could contribute to this, including more smoking, drinking and risk-taking behavior in men, as well as more differences in occupations between men and women. However, there are also a few biological explanations for the difference in lifespan between the sexes.

You probably remember from school biology that we all have two "sex chromosomes," and women usually carry XX while most men's are XY. What this nomenclature doesn't illustrate is that the Y is a stubby little thing a third the size of the X, containing dramatically fewer genes. This means that men don't have a "backup" copy of a gene if there's a problem with one of those on their single X chromosome. This is why color blindness is more common in men. There are two genes essential for color vision stored on the X chromosome: **OPN1LW** or **OPN1MW** are responsible for proteins that detect red and green light. If you're a man and one of these genes has a problem, your Y chromosome can't make up the difference and this results in an inability to distinguish red from green. Missing genetic backups have far more subtle effects when it comes to the rate of aging, but it's observed throughout the animal kingdom that whichever sex has nonmatching sex chromosomes tends to have a shorter life expectancy. In birds, for example, males have ZZ chromosomes and females ZW, and males tend to be the longer-lived sex.

It's also speculated that mitochondria may play a role in the longevity gap, thanks to the peculiar way we inherit them: exclusively from our mothers. Every one of your mitochondria is a descendant of the several hundred thousand in the egg which went on to become you, which means that a tiny fraction of your DNA (that in the mitochondria) comes not from a mixture of both your parents, but from just the maternal side. This is very strange from an evolutionary standpoint: a man with a mutation in his mitochondrial DNA that gave him a huge reproductive advantage would be unable to pass on the very mitochondrial DNA which had conferred the advantage, while a similar mitochondrial mutation in a woman would be passed on to her many daughters, and their daughters, and so on. Because of this asymmetry of mitochondrial inheritance, their evolution may improve the lot of women, without caring too much about their effect on men, resulting in a sprinkling of mitochondrial characteristics that slightly improve female fitness compared to male.

Finally, sex hormones are likely to play a role. We found out in Chapter 6 that eunuchs and castrated male prisoners outlived their contemporaries— in the case of the eunuchs, by a significant margin. If the eunuch data are to be believed, their exceptional lifespans suggest that men have the biological robustness to live longer than women, but for testosterone conspiring to kill us. (Presumably testosterone improves reproductive success at young ages, meaning that men can blame their shorter lives on sex-specific antagonistic pleiotropy.)

One piece of news that very slightly offsets the raw deal for men is that, curiously, though women live longer, they tend to do so with worse health on average. There is still some debate about the size and even the existence of this phenomenon, but perhaps the most compelling data come from centenarians: one study found that women over the age of 100 outnumber men by four to one, but that 37 percent of the centenarian men didn't suffer from any of the 14 age-related diseases in the study, compared to just 21 percent of the women.

Though being born a woman is uniquely useless advice for around half the population, it's actually the case that many of the other items on this list can be difficult or impossible for many people. For example, health problems (including those caused by advanced age) can stop people from undertaking as much exercise as they'd like; money and time constraints can make eating well more difficult for some people; planning of towns and cities can make healthy activities like commuting on foot or by bike difficult, and so on. And, though prevention is better than cure and it's never too late to start, some people are already old and unwell. Finally, advice like this won't reliably get everyone to a ripe old age in good health. It's little consolation to a clean-eating marathon runner who dies in their fifties that they were **statistically** better off than people who led less healthy lives.

For all these reasons, though dispensing health advice is hugely important, there's far more that aging biology can do to help us all lead longer, healthier lives. The next chapter will look at how we can go beyond living well individually, and help everyone live longer in good health—what governments need to do, how research needs to change, and how we all, as citizens and voters, can work to ensure a longer life for everyone thanks to biogerontology.

11

From Science to Medicine

Curing aging isn't just a scientific issue—it's going to require a shift in our politics, policies and regulations to help breakthroughs in bio-gerontology progress from research to widespread use. Given the stakes, there's a huge ethical incentive to move as quickly as we can to ensure that as many people as possible can benefit from these treatments. There's also a personal incentive for many of us alive today—if you're middle-aged or younger, lucky enough to be in good health and take care of yourself in every way you're able, the main determinant of how long you will live is progress in medicine that tackles aging.

This means that, as well as scientific discoveries, aging research needs advocacy. In this chapter we're going to look at what needs to change—from telling people about the potential for a medical revolution, to changes in policy and how we do research.

The prerequisite for all of this is a much wider understanding of the significance of the latest results in biogerontology, among everyone from scientists and doctors to politicians and the public. That's the reason I wrote this book: treating aging sounds like science fiction until you've heard about the latest developments in aging biology. That means it's often dismissed by default, covered more as a novelty than a potential reality by the media, and largely ignored by policymakers. Though there is an increasing buzz around the field now, the idea that scientists actually can slow and perhaps reverse aging in the lab still hasn't permeated popular perception. A 2013 survey found that 90 percent of Americans had heard only a little or nothing at all about treating aging—and, though it's hard to believe that things haven't improved a bit in the intervening years, it shows that we started quite recently from a pretty low base.

Scientists are also guilty of this. Because biogerontology has been a small field historically, its recognition even among biologists is surprisingly low. Aging rarely gets more than a brief mention in undergraduate lectures or textbooks, in spite of it being one of the most universal and significant processes in biology. Unaware of its importance, scientists in training do PhDs in other fields from cancer research to virology. When they come to set up labs of their own, even if they have learned about aging in the intervening time, there is little incentive to deviate from their expertise and a proven track record in their existing field. This means there are few people to lecture undergrads or accept keen PhD students, which forms something of a vicious cycle. The small size of a field of study can be a self-fulfilling prophecy.

Step one is therefore raising awareness of the amazing discoveries we've discussed so far. None of the other policy changes we need are possible without a widespread understanding that aging is something we can and should be trying to treat— and this is something all of us can participate in, whether talking to politicians, scientists, or friends and family.

The next step is that biogerontology desperately needs more funding. Research into aging is currently drastically under-resourced compared to the impact it could have on our health. Many areas of science are arguably underfunded compared to their potential impact, but aging research fares badly even compared to other scientific fields.

The U.S. is unusual in that it has a government funding body specifically dedicated to aging research (that this is unusual is of course also a problem). The National Institute on Aging, or NIA, had a budget of $2.6 billion in 2020. That's less than half of the $6.4 billion budget allocated to the

National Cancer Institute, and under 10 percent of the budget of its parent organization, the National Institutes of Health or NIH. Aging causes 85 percent of deaths in the U.S., but receives 6 percent of health research funding—substantially less than research into the diseases that aging causes.

By stark contrast, the U.S. spends $4 **trillion** every year on healthcare, a large fraction of which is spent on the chronic conditions of later life. The NIA budget is less than 0.1 percent of U.S. healthcare spending. Given that research could reduce the cost of the healthcare system with preventative treatments, this is crazy even from an economic standpoint, before we consider the huge human cost of the diseases and disabilities of old age.

The other problem is that funding streams labeled as "aging" are often used for researching diseases of aging, rather than aging itself. It's a biogerontology in-joke that NIA actually stands for National Institute for Alzheimer's because its Division of Neuroscience receives more than half of its $2.6 billion budget, while the Division of Aging Biology receives just 10 percent. And that's before we get on to the fact that "Aging Biology" concentrates on fundamental research into the mechanisms of aging, not primarily developing treatments. Fundamental research is critically important, and its discoveries underpin more practical work—but the amount of government funding trying to turn that understanding into actual treatments for aging is probably in the region of one ten-thousandth the amount of money spent on healthcare in the U.S.

The U.S. is far from alone in this. Countries around the world spend depressingly little on aging research, in spite of aging being the leading cause of disease, disability and death globally. Biogerontology is in desperate need of more money to find new ways to treat aging, and turn the ideas we already have into treatments.

Politicians shouldn't consider funding of aging research to be a cost, but an investment: one effort to calculate the benefits of anti-aging treatments found that a modest slowing of aging resulting in a 2.2-year increase in lifespan and healthspan would be worth seven trillion dollars over 50 years, just considering the health benefits to the U.S. population. The benefits to science and business would be large, too: a government that wanted to invest seriously in aging medicine would find itself at the forefront of what promises to be one of the world's largest industries, with a target market of literally every living human.

Science is cheap—even if "all" we get is a few additional healthy years of life apiece, those breakthroughs in biogerontology would be worth the comparatively tiny price tag. If we devoted $10 billion to every hallmark of aging—surely enough to make some serious progress—it would only set us back $100 billion: 2.5 percent of the U.S. annual spend on healthcare. Spread over a few years, and multiple countries, this kind of investment is surely affordable; and, if it does make serious progress against aging, we could unseat our incredible progress against infectious disease as humanity's crowning achievement. We need to petition our governments to invest more in this vital area of research. This should be an easy case to make if only politics were more rational—and the more of us who try to make the case in varied ways which will appeal to different politicians and voters, the greater our odds of success.

Though the biggest bottleneck in biogerontology is undoubtedly funding, there are also more specific ideas which are important to maximize our chances of success—changes in policy which will allow us to capitalize more quickly on scientific results and get them to patients.

The first problem, which you may remember from the Introduction, is that regulators won't currently approve a drug that treats "aging," rather than a specific disease. In the short term, this won't obstruct progress: slowing or reversing hallmarks of aging will impact the diseases they cause, and treatments can seek regulatory approval for those conditions first— for example, we've already seen how senolytics are in human trials for arthritis and lung disease, and stem cell therapies are being trialed for Parkinson's, rather than aging writ large. However, once they've proven their worth in specific conditions, the ultimate aim should be to deploy these therapies preventatively in people before they get sick—and scientists are already laying the groundwork to make this possible.

This regulatory impasse is being broken by a team of scientists, led by biogerontologist and doctor Nir Barzilai, who are conducting a revolutionary trial of a thoroughly unrevolutionary drug: metformin. Metformin is used to treat diabetes and is one of the most widely used medicines on the planet—around 80 million prescriptions for it are written every year in the U.S.. It's also got a long track record, having first been approved in the UK in 1958. This mundane molecule would be "just" an extremely safe and effective diabetes treatment, but for the unexpected positive side effects which seem to accrue to people taking it.

The most striking was a comparison of diabetic patients treated with metformin against those on another popular family of diabetes drugs called sulphonylureas, and also against a "control" group of patients who were the same age and sex but didn't have diabetes and consequently weren't taking either drug. Diabetics on metformin lived longer, not just than patients taking sulphonylureas, but they beat non-diabetics by a small margin, too—even though the patients without diabetes were healthier and less likely to be obese. There are also hints that metformin reduces the risk of cancer, heart disease and dementia, even though it's only used for treatment of diabetes. This kind of across-the-board reduction in age-related diseases and death makes it sound like this diabetes drug is having a far more fundamental effect, on the aging process itself.

Unfortunately, like the diet and exercise studies in the last chapter, this work is all observational so far. It could be, for example, that well-controlled diabetics are more resistant to other diseases of aging for some reason other than their metformin prescriptions, or perhaps they get more contact with the healthcare system so nascent problems get caught and treated earlier. What is needed is a gold-standard "randomized" trial, where whether or not you receive metformin is random, rather than based on whether or not you have diabetes.

That's the aim of the TAME trial—short for Targeting Aging with MEtformin—which will re-cruit 3,000 volunteers between 65 and 80 to test whether the drug is a true anti-aging treatment. Fifteen hundred will take the real drug and the other 1,500 will get a placebo; after five years or so, its suc-cess will be judged by whether the participants in the metformin arm of the trial get any one of a number of age-related diseases, like cancer, heart disease and dementia, later than the control group does.

The team behind TAME isn't anticipating earth-shattering results: if metformin added decades to human lifespan that would already be obvious given its widespread use. However, where metformin comes into its own is side effects—or, rather, the lack of them. After over half a century of prescribing, we know that it causes few serious problems. If you're trying to convince a risk-averse regulator to let you give pills to what they consider healthy people, then "first, do no harm" is a solid maxim. Metformin was chosen to be the first drug tested for aging itself precisely because it is pragmatic and middle-of-the-road, the drug equivalent of a family SUV with a top-notch safety record rather than a pharmaceutical supercar that might career off the racetrack. Metformin's other advantage is that it's so old, it's no longer patented. That means generic versions of it can be produced for pennies per dose, both reducing the cost of the trial and meaning that it would be practical to roll out widely if it works.

Even if the trial fails and metformin proves no better than a placebo, TAME's methodology, developed in close collaboration with the FDA, should provide an off-the-peg regulatory approach for testing future treatments. Though it would be a shame if the first large human trial of a treatment against aging gets equivocal results, denying biogerontology the opportunity to trumpet the success of its first real-world trial, this model will provide a precedent when scientists and drug companies try to get approval for the next generation of anti-aging treatments.

The other problem with anti-aging treatments is that trials take a long time, which also makes them expensive. TAME's price tag is $70 million, in spite of the fact that metformin is a very cheap drug where we can skip straight to a late-stage trial because we already know a lot about dosage and safety. On the one hand, it's a little churlish to highlight the price tag: if the trial shows that metformin works and can delay aging even slightly, it could repay this upfront cost thousands of times over. On the other, this is totally out of reach for academic scientists and a hefty sum even for a pharmaceutical company, illustrating how cost constraints can make developing treatments against aging tricky.

The expense of late-stage trials is a problem for all kinds of medical treatments but it's an especially acute one if you want to give an anti-aging drug to healthy people. A new cancer drug could cause a tumor to recede within weeks, and a longer-term trial may examine how many patients survive to five years without a relapse to demonstrate that it works—and five years is sadly long enough for many of your trial's patients to die. However, most of a cohort of relatively healthy 60-somethings given an aging treatment will still be alive after five years whether it works or not, which is obviously great news for them, but bad news for the statisticians trying to quantify the effectiveness of your new wonder drug. If you want to give your drug to healthy 30- or 40-somethings, the problem is magnified even further. Clearly, another approach is needed.

Luckily, there is a scientific solution: using "bio-markers" of aging, simple tests that can tell you someone's biological age at a moment in time. We already met one of these in Chapter 4: the "epigenetic clock," which uses chemical flags on your DNA to estimate your age (and odds of death) with unnerving accuracy.

The original epigenetic clock has now been verified many times over in different studies—in fact, it's proven so robust that labs doing completely unrelated studies on DNA methylation will quickly add up a patient's epigenetic age and check that it matches their recorded age to flag data entry errors. There are also multiple new epigenetic clocks which are less accurate predictors of chronological age— which, if you think about it, we don't really need to know, because we can deduce it using the far simpler technology of birth certificates—but are better at determining how long you might live, how long until you get cancer or heart disease, and so on.

A new version of the epigenetic clock was developed in 2018 which is a much more accurate predictor of death than the original. It also predicts cancer, Alzheimer's and, more abstractly, how many diseases someone is likely to simultaneously suffer from in the future. Unlike the original epigenetic clock, it also detects if a patient has smoked, or does so currently—a further smoking gun, if you will, suggesting that tobacco accelerates aging globally, beyond just being really bad for your lungs.

There are many other candidate biomarkers of aging, from physical examinations like grip strength and ability to stand on one leg* and lung capacity; cognitive tests and measures of vision or hearing; and more scientific-sounding ones like blood tests, brain imaging or analysis of your microbiome. There are also composite measures, which combine some or all of these to give the best possible estimate of a person's true biological age.

Perhaps simultaneously the most and least surprising aging biomarker is physical appearance. It turns out we've got reasons beyond vanity to envy those who hang on to their fresh faces as the years pass: looking young seems to mean that, biologically, you **are** young. A 2009 study asked panels of assessors to guess people's ages based on photographs of their faces. The predictions were tallied up to get an average "perceived age," which turned out to be an accurate predictor of mortality, even after accounting for chronological age. The next step is to automate this idiosyncratic and labor-intensive process with AI, which has been done with some success using both regular photographs of people and three-dimensional maps of their face shape. A team is also in the process of automating this for mice, using image recognition algorithms to deduce a mouse's biological age from a picture. This would allow researchers to evaluate anti-aging interventions in mice with just before-and-after photos.

* In an attempt to make it sound more scientific, doctors call this the "timed unipedal stance balance test."

Even though mice are much easier and cheaper to work with than humans, using them is still one of the most expensive forms of biomedical research, and this again could help cut costs and speed vital experimentation with anti-aging treatments.

Measures of biological age, or "biomarkers" of aging, are therefore incredibly useful. Instead of handing patients some pills and then putting them out to pasture for a decade, we could ask them back after a few months and see if their biological age has changed. If the ticking of their biological clock has slowed or, even better, run backward, then we can deduce that we might be onto something without needing to wait many years and check who's still alive. The other significant advantage of biomarkers is that every human or mouse in your trial can provide you with data, rather than just those who have died. This makes them much more statistically efficient, meaning you can get higher-powered studies with fewer participants.

The most important question is whether these biomarkers, good though they are at predicting risk of death or disease, are slowed or turned back by successful anti-aging interventions. Evidence is accruing that they are. The trial we mentioned in Chapter 6 that used a hormonal treatment to rejuvenate the thymus was accompanied by a reduction in subjects' epigenetic age. In mice, the ticking of their rodent epigenetic clocks is slowed by dietary restriction, treatment with rapamycin, and in mice with genes which increase lifespan. A 22-month-old mouse on DR, for example, has a biological age of just 13 months, an epigenetic manifestation of the slowing of aging expected from DR. A similar result in rhesus monkeys showed that those on DR had an epigenetic age seven years younger than those eating what they liked. There's more to be done to work out which biomarkers perform best under what circumstances, but results like these are a promising start.

If a biomarker as accurate as current epigenetic clocks was meaningfully turned back by anti-aging treatments, a study equivalent in accuracy to the TAME trial—which needs 3,000 patients, five years and tens of millions of dollars—could theoretically be completed with just a few hundred patients, two years and perhaps a few million dollars. You can view this as either a bargain-basement way to get the same results or an opportunity to test dozens of treatments (and combinations of treatments) for the same initial outlay. This is why the quest to find workable biomarkers is a particularly important subfield of biogerontology: aging has a number of underlying causes, and each one has multiple candidate treatments, so anything that makes it possible to test them rapidly and cheaply would be very welcome. Biomarkers of aging are an enabling technology which will help the wider field advance more quickly, and help us save more lives sooner.

When performing trials, it's also vital that we stop excluding the most important patients for anti-aging treatments: old people. New treatments of all kinds are rarely trialed in the elderly, even in cases where they are likely to be the primary users, because old people are "too complicated" in a variety of different ways. From a scientific perspective, you might want to test your new drug in patients who only have the particular disease you're interested in and not a laundry list of other health problems which might confuse your results. Old people are also often taking multiple different drugs for all those diseases, which may interfere with the treatment you're testing. Using younger people keeps things simple, making your results easier to interpret. There are also commercial incentives: a trial in young, fit people is more likely to show an unambiguous result which will lead to your treatment being approved. And finally, there are simple but important steps which can be taken to enable older people to participate— such as providing taxis or home visits to help those with reduced mobility—which all too often aren't taken because they're expensive and inconvenient for those running the trial.

The end result of this is that we often don't have good evidence as to whether drugs work in older people. Guidelines for some common types of medication have never been tested in the elderly, meaning, in the worst case, they could be wildly wrong. This systematic, albeit often unintentional, exclusion of old people from clinical trials needs urgent attention. Even though doctors have been saying this for decades, reality is catching up very slowly. There's a saying in pediatrics that children aren't just small adults. There should be a similar refrain in geriatrics, that old people aren't just old young people.

The same problem plays out in mouse studies. "Mouse models" of disease—which we've said before are often imperfect analogues—are particularly guilty in this regard. For example, a mouse model of Alzheimer's might contain an extra copy of the amyloid precursor protein gene, and mice could develop amyloid deposits and cognitive impairment in mouse middle age, or even in youth, unlike most human patients. This means that the mice could be relatively healthy except for the additional amyloid—great if you want to single out its effects, but not so great if you want a realistic model of human dementia.

Using older mice in experiments complicates things, just as it does in human patients, and it's also more expensive and time-consuming for the prosaic reason that you have to look after your mice for a year or two while they grow old. However, it's well known that lots of drugs which work flawlessly in mouse models fail to translate to human success. If your drug is for a disease primarily found in older patients, this is one of many possible reasons why. Given that studies in mice are often the precursor to far more expensive human trials, spending more upfront on older mice might ultimately reduce the costs of drug development by spotting duds earlier in the process.

There are some positive steps being taken. For example, some studies on vaccines have started to focus specifically on the older people who need them most. Vaccines which have a stronger "adjuvant," a substance which riles up the immune system to make it fight harder against the vaccine, or shots which simply contain more active ingredient to prod tired immune cells into action, have both been shown to work better in the elderly. Some studies show that the time of day the vaccine is administered can make a difference, with flu shots given to elderly people in the morning sometimes resulting in an enhanced immune response. More advanced ideas based on our understanding of how the immune system changes with age would be worth exploring, and it's obviously essential that these are tested in older subjects, whether mice or people.

As well as more testing in authentically old mice and people, we also need lots more work to understand the differences between old and young in more detail. For example, though we know that numbers of senescent cells increase with age, we have very few hard numbers on exactly how much, or how it varies between individuals or parts of the body. Are certain people or organs more afflicted by these cells, and do they have a greater effect in one place or another? Should this affect how we develop senolytic drugs, targeting the most consequential places first?

These important questions are finally starting to be answered with the current flurry of excitement around senolytic treatments, but it's taken a long time to get to this point. Senescent cells were first discovered in the 1960s, but it wasn't until the late 2000s that someone tried clearing them in mice to see how substantial their effect on aging organisms is. (Even this pioneering work, incidentally, was funded by cash left over from other grants after an attempt to get funding from the cash-strapped NIH was rejected. After this result was published in 2011, the scientists had more luck with their grant applications.) And this kind of work is comparatively easy to get funding for because it's interventional—simply counting the number of these cells would be tougher to raise funding for.

We're going to need a lot more of this kind of work to get a handle on what changes with age and by how much. Another example is mutations, which have been far more extensively studied in cancers than in normal aging tissue—even though cancers arise from those normal tissues. Without sequencing the DNA of non-cancerous tissue which is merely old, we run the risk of missing crucial findings which could have an impact on both cancer and the wider aging process. Similar quantitative investigation is needed for all the hallmarks of aging—changes in epigenetics, levels and modifications of proteins, numbers of cells, mitochondria, levels of signals, and so on. In the short term, this will inform our first treatments against aging. Quantification of these processes will supply biomarkers which provide an indication of whether a new drug has achieved its direct goal—for example, whether it's got rid of a significant number of mutant mitochondria, or altered the balance of age-associated signals in a meaningful way. In the longer run, this kind of data will be essential for the systems biology models we need to build of our aging bodies.

Finally, we need to prepare ourselves for what happens as evidence for the effectiveness of aging treatments starts to build up. As individuals, this is a fascinating time to be alive. As we rapidly learn more about treatments which could intervene in aging, it's natural to wonder at what point the risks and benefits of a new treatment are well enough characterized to consider taking it. Our current paradigm of medical research works on the precautionary principle, meaning that pharmaceutical companies and regulators expend huge efforts to make absolutely sure that new treatments are completely safe before rolling them out for widespread use. This sounds sensibly cautious, but it neglects that doing nothing can sometimes come with greater risks than doing something, even if that something is not 100 percent certain to be safe. This problem of balancing risk becomes particularly acute with aging treatments which, ultimately, we hope will be deployed preventatively—perhaps to a significant fraction of the population—before people become unwell.

We're all going to need help navigating this new paradigm in medicine, because the calculus involved will be quite different from current drugs. Would you be happy to start taking pills in your forties to slow down your own aging? How much evidence would you need to see before it seemed like the right call? The fact that we might be committing to a course of treatment while not suffering from any disease without knowing its lifelong effects with certainty is a challenge for regulators and individuals alike. But, equally clearly, we can't afford to wait for 50-year-long trials to churn out definitive answers when acting sooner has the potential to save and improve millions or billions of lives.

At the same time, we need to keep charlatans at bay. Anti-aging medicine already has a checkered history of quacks peddling immortality with un-evidenced treatments from potions and elixirs to, astonishingly, surgical implantation of animal testicles. It's hard for non-experts to weigh the evidence for a treatment, or even to know whether the treatment **is** chemically or biologically what it claims to be. Smart regulators and solid public information are required to make sure that people don't waste money or get hurt.

Finally, we should think seriously about how we might standardize protocols and capture data from people who are already self-experimenting with these treatments. It only takes a quick Web search to find groups of people speculatively taking metformin for its anti-aging effects, perhaps by asking their doctor to prescribe it to them in spite of not having diabetes. At the other end of the spectrum of risk to reward, there's a case of a biotech CEO going to a Colombian clinic to give herself untested, unregulated telomerase gene therapy. There's clearly a huge appetite for this kind of experimentation—and some professional oversight could make these homespun single-person clinical trials both safer for participants and more useful for all of us.

If people are going to do it anyway, it would be a terrible waste if their experiments are atomized, idiosyncratic and uncontrolled: we might never find out the results and, even if we did, infinite variations in practical protocol will mean we never really know if their efforts helped them to live longer or not. By introducing at least some of the rigor of a conventional clinical trial, ensuring that participants are using the same dose of the same drug which is what it says it is, these self-experiments could not only be made safer, but provide far more generalizable knowledge about which interventions work and which don't.

Implementing this would be challenging due to the need to both assess and communicate risk and uncertainty very carefully. If I were a 65-year-old already contemplating a somewhat uncertain therapy, my willingness would be increased if I not only had the chance to live a bit longer in good health, but my modest gamble could also contribute to a better understanding of aging for the generations who follow.

Clearly, the success of biogerontology depends on more than just science. The essential prerequisites are a much higher profile and far more funding for aging research. It will need to clear hurdles in policy and regulation and, perhaps most important of all, biogerontology needs to go mainstream, with widespread understanding and discussion of its potential among scientists, policymakers and the public.

I hope this book has convinced you that it's time for a mission-driven medical moonshot—a massively funded, international program of research to intervene in the aging process. It would be bafflingly unlucky if we didn't find a few new and innovative ways to improve human health, and there's potential for rewards that are far greater still.

Even if we aren't fortunate enough to be the first generation of ageless humans, longer, healthier lives will bring enormous benefits, for us and for every future generation.

Every day we bring forward a cure for aging, we save 100,000 lives. We know it's scientifically possible. It's now up to all of us to meet the defining humanitarian challenge of our time.

Acknowledgments

The ideas in this book have been my passion for almost a decade, after discovering aging biology as I was finishing my PhD. I would like to thank everyone who has made this book possible and I hope that, with their help, I have done justice to what I genuinely believe are the most important scientific ideas of our age.

I would like to start by thanking the generations of researchers upon whose work this book is based, and scientists past and present whose names or work I have not been able to mention specifically in this brief tour of biogerontology. You will find some of their names in the references that follow these acknowledgments, but there are of course many more besides. Without the historical and ongoing efforts of the scientific community, there would be no exciting results to report, and we would not find ourselves at this pivotal time in the history of medicine.

I have been humbled by the scientists and others who have given so generously of their time to patiently answer my (often naïve!) questions and read through drafts of this manuscript. In the following lists, names appear in a random order.

First, it has been a huge honor to speak with some truly incredible researchers and others, all of whom have expanded my horizons, brought the ideas I read about in the literature to life and, in some cases, overturned what I'd read with cutting-edge results of their own. Thanks to Nick Lane, Desmond Tobin, Jon Houseley, João Pedro de Magalhães, Adam Rolt, Melinda Duer, Graham Ruby, Mike Philpott, Aubrey de Grey, Linda Partridge, David Gems, Sebastian Aguiar, Jim Mellon, Judith Campisi, Wolf Reik and Anders Sandberg.

Second, I would like to thank those who read and commented on drafts of sections of the book. The text is much improved for their thoughtful reviews. Thanks to Jonathan Slack, Hannah Ritchie, Robert J. Shmookler Reis and María Blasco.

Third, special thanks to those who both spoke to me and found the energy to read and comment on drafts as well! Thanks to Hannah Walters, Anna Poetsch, Alejandro Ocampo, Jonathan Clark, Iñigo Martincorena, Adrian Liston, Richard Faragher, Nir

Barzilai, Irina and Mike Conboy, Didac Carmona-Gutierrez, João Passos and Michelle Linterman.

I would also like to thank the many people not listed here, from biologists to historians, doctors to actuaries, who replied to my messages or spoke to me briefly to give more details behind work I'd read—from the pedantic and statistical, to the frivolous, such as my attempt to track down whether the bacterium behind rapamycin was found **actually under** one of the famous statues on Easter Island as some accounts imply (after disproportionate effort, inconclusive), or asking them to dig up decade-old data so I could accurately report on what day the last worm in an experiment died (see Chapter 3).

Lastly, I would like to thank three people who were kind enough to read a draft of the entire book. Thank you so much to my friends Tom Fuller and Maya Evans, and especially biogerontologist Lynne Cox, whose fresh eyes and biological insight improved the manuscript in its final stages.

Writing a factual book is never a solo enterprise, and, thanks to all of these people, my writing is more accurate, more interesting and more complete. Any errors or omissions are my own.

I would like to thank the Francis Crick Institute for allowing me to continue as a visiting researcher, allowing me to retain access to the scientific literature which underpins this book, in particular to Nick Luscombe for giving a physicist a chance to work in biology, and to the whole Bioinformatics and

Computational Biology Lab for helping give me the grounding without which I would not have been able to write it.

I am also hugely indebted to my editors, Alexis Kirschbaum, Kristine Puopolo and Jasmine Horsey, for their faith in my writing, for finding the book you've just read hidden in my first draft and for making the editing process thoroughly enjoyable. This book wouldn't have been possible without the brilliant work of my agent, Chris Wellbelove, and the rest of the team at Aitken Alexander, who guided it masterfully from inception to publication. Many thanks also to everyone else at my UK and U.S. publishers, Bloomsbury and Doubleday, for all of their work proofreading, typesetting, designing beautiful covers, marketing the book, and much more besides.

Finally, I would like to thank my wife, Tran Nguyen, who has helped shape every part of this book with many hours of discussion, comments on numerous drafts and as an invaluable in-house source of medical expertise, and whose many stories of her older patients struggling with ill health reminded me every day why it was so important to keep writing.

Notes and Bibliography

These notes list some of the sources I used while writing. I've tried to provide a citation for all the important facts and figures in the text, using sources that are free to read where possible, and giving preference to more accessibly written studies—though obviously scientific papers can sometimes be quite heavy going. Occasionally I've referenced particularly good popular science articles, books and videos that often go into greater depth on a specific topic than I've had space to in this book. I also haven't always referenced the primary source where another provides a more readable overview.

All citations are accompanied by a short link, starting with ageless.link/ and followed by a unique code made up of six letters and numbers (e.g., ageless .link/m3gh76). If you type this into a Web browser, it will take you to the reference, occasionally with additional information. Visit ageless.link/references for more information.

INTRODUCTION

2 **"negligible senescence"—a negligible loss of capacity as they age:** Owen R. Jones and James W. Vaupel, "Senescence is not inevitable," **Biogerontology** 18, 965–71 (2017). DOI: 10.1007/s10522-017-9727-3 ageless.link/i3hrtb.

3 **We, by contrast . . . every eight years:** Calculation based on Human Mortality Database data. You can read about how this was done at ageless.link/e7ywum.

5 **An 80-year-old . . . to get heart disease:** Calculations based on World Health Organization Global Burden of Disease (WHO GBD) statistics. You can read about how this was done at ageless .link/cxspho.

5 **Having high blood pressure . . . your risk by ten:** Emelia J. Benjamin et al., "Heart disease and stroke statistics—2017 update: A report from the American Heart Association," **Circulation** 135, e146–e603 (2017). DOI: 10.1161/CIR.0000000000000485 ageless.link/wxyygy.

5 **The average 80-year-old . . . types of medication:** According to the first two studies here, an 80-year-old suffers from an average of three "morbidities" and five to ten "diagnoses." The difference between them is methodological: how many "diseases" someone suffers from depends on what your

threshold is, so I said "around five" in the text. The third paper estimates the number of drugs, or "polypharmacy." Thanks to Bruce Guthrie for helping me understand the nuances.

Karen Barnett et al., "Epidemiology of multimorbidity and implications for health care, research, and medical education: A cross-sectional study," **Lancet** 380, 37–43 (2012). DOI: 10.1016/S0140-6736(12)60240-2 ageless.link/itozkk.

Quintí Foguet-Boreu et al., "Multimorbidity patterns in elderly primary health care patients in a south Mediterranean European region: A cluster analysis," **PLoS One** 10, e0141155 (2015). DOI: 10.1371/journal.pone.0141155 ageless.link/e4q6vg.

Bruce Guthrie et al., "The rising tide of polypharmacy and drug-drug interactions: Population database analysis 1995–2010," **BMC Med.** 13, 74 (2015). DOI: 10.1186/s12916-015-0322-7 ageless.link/7enffk.

7 **Over 100,000 of them die because of aging:** This was calculated using WHO GBD statistics. You can read about the calculations at ageless.link/hbzze7.

8 **Research shows that . . . it will happen to <u>them</u>:** In this study, 250 students were asked to rate their chances of experiencing positive or negative life events, like owning a home, living past 80, getting lung cancer or having a heart attack. They consistently overrate their own chances of positive experiences, while thinking bad things are more likely to happen to others.

Neil D. Weinstein, "Unrealistic optimism about future life events," **J. Pers. Soc. Psychol.** 39, 806–20 (1980). DOI: 10.1037/0022-3514.39.5.806 ageless.link/pouimx.

9 **a U.S. survey found . . . age of 63 themselves:** "Who are family caregivers?" (American Psychological Association, 2011) ageless.link/ufntz3.

9 **age-related diseases . . . around the world:** This was calculated using WHO GBD statistics. You can see the calculation at ageless.link/hbzze7.

10 **surveys show . . . state of the world:** Our general state of pessimism about the world—from life expectancy, to education, to vaccine provision—is captured beautifully by the brilliant Hans Rosling's Ignorance Survey.

Hans Rosling, "Highlights from ignorance survey in the UK" (Gapminder Foundation, 2013) ageless.link/4qppjz.

10 **a breakthrough that changed scientific history:** Clive M. McCay, Mary F. Crowell and L. A. Maynard, "The effect of retarded growth upon the length of life span and upon the ultimate body size," **J. Nutr.** 10, 63–79 (1935). DOI: 10.1093/jn/10.1.63 ageless.link/ovmys4.

13 **diversity of life on Earth:** This fun and accessibly written article explores some ways, from aging to regenerative medicine, that animals can inform biomedical science: João Pedro de Magalhães, "The big, the bad and the ugly: Extreme animals as inspiration for biomedical research," **EMBO Rep.** 16,

771–6 (2015). DOI: 10.15252/embr.201540606 ageless.link/qjy7oo.

15 **In 2011 . . . and extend lifespan:** Darren J. Baker et al., "Clearance of p16Ink4a-positive senescent cells delays aging-associated disorders," **Nature** 479, 232–6 (2011). DOI: 10.1038/nature10600 ageless.link/qqyqtf.

15 **By 2018 . . . trials in people:** Jamie N. Justice et al., "Senolytics in idiopathic pulmonary fibrosis: Results from a first-in-human, open-label, pilot study," **EBioMedicine** 40, 554–63 (2019). DOI: 10.1016/j.ebiom.2018.12.052 ageless.link/phgw6r.

17 **cancer—currently the leading cause of death:** What exactly is the leading cause of death depends on your classification system and the region of the world. This claim refers specifically to the World Bank "High-Income Countries," and is based on my own aggregation of World Health Organization data which groups all cancer types together. If you split cancer into types, aggregate strokes, heart attacks and so on into "cardiovascular disease," or look in specific countries, your results may differ. You can see these figures (which I have used when ranking causes of death throughout the book) at ageless.link/a6rv67.

17 **a complete cure for cancer . . . three years to life expectancy:** G. D. Wang et al., "Potential gains in life expectancy from reducing heart disease, cancer, Alzheimer's disease, kidney disease or HIV/AIDS as major causes of death in the USA,"

Public Health 127, 348–56 (2013). DOI: 10.1016/j.puhe.2013.01.005 ageless.link/c7bwrm.

18 **the benefits will far outweigh the costs:** The ethics of curing aging is obviously far too large a subject to cover in a couple of paragraphs. Further reading can be found at ageless.link/ethics.

20 **People born with Hutchinson–Gilford progeria:** Leslie B. Gordon, W. Ted Brown and Francis S. Collins, "Hutchinson–Gilford progeria syndrome," in **GeneReviews** (ed. Margaret P. Adam et al.) (Seattle, WA: University of Washington, Seattle, 2003) ageless.link/ixa4uj.

20 **Another related disease, Werner syndrome:** Junko Oshima, George M. Martin and Fuki M. Hisama, "Werner syndrome," in **GeneReviews** (ed. Margaret P. Adam et al.) (Seattle, WA: University of Washington, Seattle, 2002) ageless.link/edpehq.

1. THE AGE OF AGING

26 **life expectancy . . . 30 and 35 years:** The authors in this paper estimate an expected age at death of 54 for groups of human foragers who made it to age 15: Hillard Kaplan et al., "A theory of human life history evolution: Diet, intelligence, and longevity," **Evolutionary Anthropology: Issues, News, and Reviews** 9, 156–85 (2000). DOI: 10.1002/1520-6505(2000)9:4<156::AID -EVAN5>3.0.CO;2-7 ageless.link/n4irx9.

A readable summary of child mortality in the past can be found in: Max Roser, "Mortality in the past—around half died as children," **Our World in Data** (2019) ageless.link/hrw43b.

The estimates of child mortality in that article are from Anthony A. Volk and Jeremy A. Atkinson, "Infant and child death in the human environment of evolutionary adaptation," **Evol. Hum. Behav.** 34, 182–92 (2013). DOI: 10.1016/j .evolhumbehav.2012.11.007 ageless.link/eawqcs.

Combining these sources, you arrive at a life expectancy at birth of 30 to 35 years.

28 **some of the earliest philosophers grappled with aging and death:** Gareth B. Matthews, "Death in Socrates, Plato, and Aristotle," in **The Oxford Handbook of Philosophy of Death** (ed. Ben Bradley, Fred Feldman and Jens Johansson) (Oxford University Press, 2012). DOI: 10.1093/ oxfordhb/9780195388923.013.BCE; Adam Woodcox, "Aristotle's theory of aging," **Cahiers Des Études Anciennes** LV | 2018, 65–78 (2018) ageless.link/vdhzmr.

29 **the UK and Sweden . . . early nineteenth century:** Max Roser, Esteban Ortiz-Ospina and Hannah Ritchie, "Life expectancy," **Our World in Data** (2013) ageless.link/mcviaq.

29 **since 1840, with clockwork regularity:** This paper on the historical rise of life expectancy is a provocative and entertaining read: Jim Oeppen and James W. Vaupel, "Broken limits to life expectancy," **Science** 296, 1029–31 (2002). DOI: 10.1126/science.1069675 ageless.link/ gnjkds.

30 **A 20-year-old today . . . of having a living mother:** This statistic isn't quite as impressive as it sounds because you have two grandmothers but only one mother, giving you two statistical chances to have one still alive—but the numbers underlying it actually are (almost) as impressive as it sounds, because each of your grandmothers individually are roughly as likely to be alive when you're 20 as your mother would have been in the 1800s.

P. Uhlenberg, "Mortality decline in the twentieth century and supply of kin over the life course," **Gerontologist** 36, 681–5 (1996). DOI: 10.1093/geront/36.5.681 ageless.link/jyfyrp.

31 **50 to 100 million died . . . of flu viruses:** Max Roser, "The Spanish flu (1918–20): The global impact of the largest influenza pandemic in history," **Our World in Data** (2020) ageless.link/odbnbx.

32 **By 1950, Norwegians . . . could expect to live beyond 70:** Oeppen and Vaupel, 2002 ageless.link/gnjkds.

34 **in the 1960s . . . more than half a pack a day: Fifty Years of Change: 1964–2014** (Centers for Disease Control and Prevention (US), 2014). ageless.link/nqdh7w. This paper is a very readable summary of the epidemiological evidence against smoking. It's full of similarly shocking statistics, and makes for sobering reading.

Richard Peto et al., "Smoking, smoking cessation, and lung cancer in the UK since

1950: Combination of national statistics with two case-control studies," **BMJ** 321, 323–9 (2000). DOI: 10.1136/bmj.321.7257.323 ageless.link/bukftz.

34 **a sixth of all deaths . . . were attributable to tobacco:** Calculated based on data from deathsfromsmoking.net ageless.link/di96gq.

34 **100 million deaths from smoking in the twentieth century:** Prabhat Jha, "Avoidable global cancer deaths and total deaths from smoking," **Nat. Rev. Cancer** 9, 655–64 (2009). DOI: 10.1038/nrc2703 ageless.link/fjnhnq.

34 **A study looking at . . . 1991 and 2011:** Carol Jagger et al., "A comparison of health expectancies over two decades in England: Results of the cognitive function and aging study I and II," **Lancet** 387, 779–86 (2016). DOI: 10.1016/S0140-6736(15)00947-2 ageless.link/fvztx9.

35 **the fraction of over-85s in the U.S.:** Kenneth G. Manton, Xiliang Gu and Vicki L. Lamb, "Change in chronic disability from 1982 to 2004/2005 as measured by long-term changes in function and health in the U.S. Elderly population," **Proc. Natl. Acad. Sci. U.S.A.** 103, 18374–9 (2006). DOI: 10.1073/pnas.0608483103 ageless.link/7m9pwk.

35 **severe disability is dropping, minor disability:** James W. Vaupel, "Biodemography of human aging," **Nature** 464, 536–42 (2010). DOI: 10.1038/nature08984 ageless.link/4wzcxd.

38 **one in six of the global population:** "World

population aging 2019 highlights" (United Nations, Department of Economic and Social Affairs, Population Division, 2019) ageless.link/uemmm6.

38 **the number of people aged 100 or over:** "World population prospects 2019, online edition. Rev. 1" (United Nations, Department of Economic and Social Affairs, Population Division, 2019) ageless .link/smxq93.

39 **population aging . . . developing world:** "World population aging 2015" (United Nations, Department of Economic and Social Affairs, Population Division, 2015) ageless.link/n47kou.

39 **the age at which . . . almost a century:** "History of pensions: A brief guide," **BBC News** (2005) ageless.link/nygivk.

   Jonathan Cribb and Carl Emmerson, "Retiring at 65 no more? The increase in the state pension age to 66 for men and women" (Institute for Fiscal Studies, 2019) ageless.link/cm3yqi.

41 **It seems unlikely that three stages: The 100-Year Life** is a great exploration of how our lives will need to change as lifespans extend. Lynda Gratton and Andrew Scott, **The 100-Year Life: Living and Working in an Age of Longevity** (Bloomsbury Publishing, 2020) ageless.link/9aeoey.

41 **the average 80-year-old costs the healthcare systems:** Paul Johnson et al., **Securing the Future: Funding Health and Social Care to the 2030s** (The IFS, 2018) ageless.link/up4igu.

   Bradley Sawyer and Gary Claxton, "How do

health expenditures vary across the population?," **Peterson–Kaiser Health System Tracker** (2019) ageless.link/4b3ek3.

41 **typical rich countries . . . spend roughly 10 percent of GDP:** "Current health expenditure (% of GDP)," **World Health Organization Global Health Expenditure Database** ageless.link/jhkq7u.

42 **the indirect costs of diseases . . . exceed the direct ones:** Ramon Luengo-Fernandez et al., "Economic burden of cancer across the European Union: A population-based cost analysis," **Lancet Oncol.** 14, 1165–74 (2013). DOI: 10.1016/S1470-2045 (13)70442-X ageless.link/4qenyb.

Raphael Wittenberg et al., "Projections of care for older people with dementia in England: 2015 to 2040," **Age and Aging** 49, 264–9 (2020). DOI: 10.1093/aging/afz154 ageless.link/cfzxs4.

42 **unpaid care alone . . . healthcare budget:** Sue Yeandle and Lisa Buckner, "Valuing Carers 2015" (Carers UK, 2015) ageless.link/bmn3s3.

43 **XT9T, for conditions that are "aging-related":** The Lancet Diabetes Endocrinology, "Opening the door to treating aging as a disease," **Lancet Diabetes Endocrinol** 6, 587 (2018). DOI: 10.1016/S2213-8587(18)30214-6 ageless.link/yxq7dd.

Khaltourina Daria et al., "Aging fits the disease criteria of the international classification of diseases," **Mech. Aging Dev.** 111230 (2020). DOI: 10.1016/j.mad.2020.111230 ageless.link/qvr6q9.

44 **most babies lucky enough . . . celebrate their 100th:** Vaupel, 2010 ageless.link/4wzcxd.

44 **One study examined . . . human life expectancy:** Oeppen and Vaupel, 2002 ageless.link/gnjkds.

2. ON THE ORIGIN OF AGING

A readable summary of the modern evolutionary understanding of aging which goes beyond the framework set out in this chapter can be found in: Thomas Flatt and Linda Partridge, "Horizons in the evolution of aging," **BMC Biol.** 16, 93 (2018). DOI: 10.1186/s12915-018-0562-z ageless.link/ktangr.

50 **the longest-lived vertebrate . . . 400 years old:** Julius Nielsen et al., "Eye lens radiocarbon reveals centuries of longevity in the Greenland shark (**Somniosus microcephalus**)," **Science** 353, 702–4 (2016). DOI: 10.1126/science.aaf1703 ageless.link/x9mkhj.

52 **the first evolutionary theory of aging:** Michael R. Rose et al., "Evolution of aging since Darwin," **J. Genet.** 87, 363–71 (2008). DOI: 10.1007/s12041-008-0059-6 ageless.link/zasohq.

57 **we each carry 50 to 100 variations:** Catarina D. Campbell and Evan E. Eichler, "Properties and rates of germline mutations in humans," **Trends Genet.** 29, 575–84 (2013). DOI: 10.1016/j.tig.2013.04.005 ageless.link/ag4z34.

61 **Thus, the second idea . . . "antagonistic pleiotropy":** This is Williams' original paper, proposing the idea: George C. Williams, "Pleiotropy,

natural selection, and the evolution of senescence," **Evolution** 11, 398–411 (1957). DOI: 10.1111/ j.1558-5646.1957.tb02911.x ageless.link/pjritd.

This more modern paper is an interesting review of the evidence for specific examples of antagonistic pleiotropy, in the lab and in the wild: Steven N. Austad and Jessica M. Hoffman, "Is antagonistic pleiotropy ubiquitous in aging biology?," **Evol. Med. Public Health** 2018, 287–94 (2018). DOI: 10.1093/emph/eoy033 ageless.link/ 9pftdn.

62 **our third and final . . . "disposable soma theory":** T. B. Kirkwood, "Evolution of aging," **Nature** 270, 301–4 (1977). DOI: 10.1038/270301a0 ageless .link/kzwpbf.

64 **mice have litters . . . once a month:** Mouse breeding and lifespan statistics are from the excellent **AnAge** database, which curates longevity and associated data for thousands of species.

"House mouse (**Mus musculus**)," **AnAge: The animal aging and longevity database** (2017) ageless.link/z334yj.

65 **bowhead whale . . . 211 years old:** "Bowhead whale (**Balaena mysticetus**)," **AnAge: The animal aging and longevity database** (2017) ageless.link/ 7qej3n.

65 **a remarkable tale of a whale that got away:** Amanda Leigh Haag, "Patented harpoon pins down whale age," **Nature News** (2007). DOI: 10.1038/news070618-6 ageless.link/ teouks.

66 **mouse-eared bat . . . 37 when it died:** "Mouse-eared bat (**Myotis myotis**)," **AnAge: The animal aging and longevity database** (2017) ageless.link /uxa3ng.

67 **they too can live over 30 years:** Flatt and Partridge, 2018 ageless.link/ktangr.

68 **the verified chimp longevity champion:** "Chimpanzee (**Pan troglodytes**)," **AnAge: The animal aging and longevity database** (2017) ageless.link/sbc7fh.

69 **female fish . . . as they age:** Flatt and Partridge, 2018 ageless.link/ktangr.

69 **underwater matriarchs are known as BOFFFFs:** Mark A. Hixon, Darren W. Johnson and Susan M. Sogard, "BOFFFFs: On the importance of conserving old-growth age structure in fishery populations," **ICES J. Mar. Sci.** 71, 2171–85 (2014). DOI: 10.1093/icesjms/fst200 ageless.link/9k6r3u.

70 **rougheye rockfish . . . 205 years old:** "Rougheye rockfish (**Sebastes aleutianus**)," **AnAge: The animal aging and longevity database** (2017) ageless.link/pynfqt.

71 **the result is that they don't seem to age:** This paper, published in 2003, documents the original observations showing that these two species of turtle don't age: Justin D. Congdon et al., "Testing hypotheses of aging in long-lived painted turtles (**Chrysemys picta**)," **Exp. Gerontol.** 38, 765–72 (2003). DOI: 10.1016/s0531-5565(03)00106-2 ageless.link/9a7ewp.

However, 13 years later, another study looking

at a different population of painted turtles found that they do senesce (albeit slowly). As ever, the devil is in the details: proposed explanations include methodological differences, or increased extrinsic mortality in the population examined in this study (thought, incidentally, to be caused by humans—boats when in the water, and cars on the land). Tortoise demography is a niche field of research, and these projects take decades, so the debate about how tortoise mortality changes with age won't be settled any time soon. However, regardless of the outcome, the key point remains: there is an incredible diversity of life courses, and no law of nature precludes negligible senescence. (And, if the extrinsic mortality differences prove to be the clinching difference here, this would further underscore our theories about how aging evolves.)

Daniel A. Warner et al., "Decades of field data reveal that turtles senesce in the wild," **Proc. Natl. Acad. Sci. U.S.A.** 113, 6502–7 (2016). DOI: 10.1073/pnas.1600035113 ageless.link/tzrfyn.

72 **hydra . . . 1,000 years old:** This paper brings together a wealth of life course data for a huge range of species, and shows the incredible diversity of senescence across living organisms—rapid, negligible, negative, and some slightly stranger shapes. Truly evolution will maximize for reproductive success, doing whatever it takes to optimize age-related mortality to get there.

Owen R. Jones et al., "Diversity of aging across

the tree of life," **Nature** 505, 169–73 (2014). DOI: 10.1038/nature12789 ageless.link/de3y4w.

72 **Hydra violate . . . disposable soma theory:** As well as discussing this proposed mechanism for negligible senescence in hydra, this is also a very readable review of evolutionary theories of aging: T. B. Kirkwood and S. N. Austad, "Why do we age?," **Nature** 408, 233–8 (2000). DOI: 10.1038/35041682 ageless.link/ebdxpa.

72 **The tree . . . estimated 4,850 years old:** "OLDLIST, a database of old trees" (Rocky Mountain Tree-Ring Research) ageless.link/ xdrnrq.

   A fascinating feature about bristlecone pine research can be found at Alex Ross, "The past and the future of the earth's oldest trees," **New Yorker** (2020) ageless.link/x9r73z.

73 **one theory is . . . competition for space:** Robert M. Seymour and C. Patrick Doncaster, "Density dependence triggers runaway selection of reduced senescence," **PLoS Comput. Biol.** 3, e256 (2007). DOI: 10.1371/journal.pcbi.0030256 ageless.link/ ikq4ry.

74 **the possibility of <u>negative</u> senescence:** James W. Vaupel et al., "The case for negative senescence," **Theor. Popul. Biol.** 65, 339–51 (2004). DOI: 10.1016/j.tpb.2003.12.003 ageless.link/ fnujcb.

74 **the best data we have on the desert tortoise:** Jones et al., 2014 ageless.link/de3y4w.

3. THE BIRTH OF BIOGERONTOLOGY

An entertaining account of dietary restriction, early nematode worm research and some more recent developments in biogerontology can be found in David Stipp, **The Youth Pill: Scientists at the Brink of an Anti-Aging Revolution** (Current Publishing, 2010) ageless.link/7oqrph.

Live, fast, die old

80 **Clive McCay . . . the first meticulous experiment:** Clive M. McCay and Mary F. Crowell, "Prolonging the life span," **Sci. Mon.** 39, 405–14 (1934) ageless.link/is3i7p.

Additional background on McCay's career and the development of dietary restriction research more generally can be found in these two articles: Hyung Wook Park, "Longevity, aging, and caloric restriction: Clive Maine McCay and the construction of a multidisciplinary research program," **Hist. Stud. Nat. Sci.** 40, 79–124 (2010). DOI: 10.1525/hsns.2010.40.1.79 ageless.link/dggrds.

Roger B. McDonald and Jon J. Ramsey, "Honoring Clive McCay and 75 years of calorie restriction research," **J. Nutr.** 140, 1205–10 (2010). DOI: 10.3945/jn.110.122804 ageless.link/hqegja.

80 **the last one standing died at 1,321 days:** McCay, Crowell, and Maynard, 1935 ageless.link/sdakif.

82 **modern research . . . the calories themselves:** A particularly intriguing set of experiments

showed that fruit fly health and lifespan can be optimized by feeding them amino acids tailored to how often the three-letter code for that amino acid (a codon) appears in their DNA: Matthew D. W. Piper et al., "Matching dietary amino acid balance to the in silico-translated exome optimizes growth and reproduction without cost to lifespan," **Cell Metab.** 25, 610–21 (2017). DOI: 10.1016/j.cmet.2017.02.005 ageless.link/9gscb6.

83 **Single-celled yeast . . . by 300 percent:** Most lifespan extensions from DR were found in Table 1 of this review article: William Mair and Andrew Dillin, "Aging and survival: The genetics of life span extension by dietary restriction," **Annu. Rev. Biochem.** 77, 727–54 (2008). DOI: 10.1146/annurev.biochem.77.061206.171059 ageless.link/mm4wvt.

The results for lemurs were reported in this paper: Fabien Pifferi et al., "Caloric restriction increases lifespan but affects brain integrity in grey mouse lemur primates," **Communications Biology** 1, 30 (2018). DOI: 10.1038/s42003-018-0024-8 ageless.link/g6rytx.

84 **two studies in . . . rhesus macaques:** These two articles try to reconcile the differences between the studies when substantial results were first available for both in 2012: Steven N. Austad, "Aging: Mixed results for dieting monkeys," **Nature** 489, 210–11 (2012). DOI: 10.1038/nature11484 ageless.link/jxcnjr.

Bill Gifford, "Long-awaited monkey study casts

doubt on longevity diet," **Slate** magazine, 2012 ageless.link/6mrygw.

This more recent paper tries to draw together results from both the NIA and Wisconsin studies including more recent data, with more optimistic conclusions: Julie A. Mattison et al., "Caloric restriction improves health and survival of rhesus monkeys," **Nat. Commun.** 8, 14063 (2017). DOI: 10.1038/ncomms14063 ageless.link/d3ntbn.

84 **short-term markers . . . seem to be improved:** William E. Kraus et al., "2 years of calorie restriction and cardiometabolic risk (CALERIE): Exploratory outcomes of a multicenter, phase 2, randomized controlled trial," **Lancet Diabetes Endocrinol** 7, 673–83 (2019). DOI: 10.1016/S2213-8587(19)30151-2 ageless.link/deo9cn.

86 **The most popular idea:** Flatt and Partridge, 2018 ageless.link/7itruu.

The worm has turned 150

88 **a compost heap in Bristol:** Mark G. Sterken et al., "The laboratory domestication of **Caenorhabditis elegans**," **Trends Genet.** 31, 224–31 (2015). DOI: 10.1016/j.tig.2015.02.009 ageless.link/hkjgme.

89 **a project called OpenWorm:** Gopal P. Sarma et al., "OpenWorm: Overview and recent advances in integrative biological simulation of **Caenorhabditis elegans**," **Philos. Trans. R. Soc.**

Lond. B Biol. Sci. 373 (2018). DOI: 10.1098/rstb.2017.0382 ageless.link/96ocjy.

90   **In 1983, scientist Michael Klass:** The story of the discovery of longevity mutants in **C. elegans,** starting with Klass, is told autobiographically in this article by Cynthia Kenyon, another protagonist who appears later in this chapter: Cynthia Kenyon, "The first long-lived mutants: Discovery of the insulin/IGF-1 pathway for aging," **Philos. Trans. R. Soc. Lond. B Biol. Sci.** 366, 9–16 (2011). DOI: 10.1098/rstb.2010.0276 ageless.link/oaqt67.

92   **the worm-children . . . had normal lifespans:** D. B. Friedman and T. E. Johnson, "Three mutants that extend both mean and maximum life span of the nematode," **Caenorhabditis elegans,** define the **age-1** gene," **J. Gerontol.** 43, B102–9 (1988) ageless.link/ngrarj.

93   **Johnson had just confirmed them:** Kenyon, 2011 ageless.link/oaqt67.

94   **daf-2 mutants lived twice as long:** C. Kenyon et al., "A **C. elegans** mutant that lives twice as long as wild type," **Nature** 366, 461–4 (1993). DOI: 10.1038/366461a0 ageless.link/yxdvef.

95   **Worms carrying [age-1 (mg44)] live . . . 150 days:** Srinivas Ayyadevara et al., "Remarkable longevity and stress resistance of nematode PI3K-null mutants," **Aging Cell** 7, 13–22 (2008). DOI: 10.1111/j.1474-9726.2007.00348.x ageless.link/3faznm.

96   **Cynthia Kenyon . . . "the grim reaper":** Kenyon calls **daf-2** "the grim reaper" in her TED Talk,

which is a nice brief summary of her work: Cynthia Kenyon, "Experiments that hint of longer lives" (TEDGlobal, 2011) ageless.link/nzovin.

98 **These include the Laron mouse:** Holly M. Brown-Borg and Andrzej Bartke, "GH and IGF1: Roles in energy metabolism of long-living GH mutant mice," **J. Gerontol. A Biol. Sci. Med. Sci.** 67, 652–60 (2012). DOI: 10.1093/gerona/gls086 ageless.link/ac37ax.

99 **a condition . . . Laron syndrome:** Jaime Guevara-Aguirre et al., "Growth hormone receptor deficiency is associated with a major reduction in pro-aging signaling, cancer, and diabetes in humans," **Sci. Transl. Med.** 3, 70ra13 (2011). DOI: 10.1126/scitranslmed.3001845 ageless.link/vptky6.

Nicholas Wade, "Ecuadorean villagers may hold secret to longevity," **New York Times** (11 February 2011) ageless.link/vb7nvm.

100 **N2s . . . outcompete their mutant cohabitants:** Austad and Hoffman, 2018 ageless.link/r4nh7k.

100 **non mutant worms actually lived longer:** Wayne A. Van Voorhies, Jacqueline Fuchs and Stephen Thomas, "The longevity of **Caenorhabditis elegans** in soil," **Biol. Lett.** 1, 247–9 (2005). DOI: 10.1098/rsbl.2004.0278 ageless.link/zdafyk.

102 **over 1,000 genes which can increase lifespan:** Counts of longevity-associated genes were retrieved from the GenAge database of aging-related genes. ageless.link/ndu3qk.

4. WHY WE AGE

This chapter is largely structured around a paper called "The hallmarks of aging." There is a huge amount of information in the article, but it's a very dense read! Carlos López-Otín et al., "The hallmarks of aging," **Cell** 153, 1194–1217 (2013). DOI: 10.1016/j.cell.2013.05.039 ageless.link/m3gh76.

An accessible account of the hallmarks is also given in this presentation to a UK government inquiry into aging, of which you can watch a video or read a transcript. Other hearings and written evidence to this inquiry are interesting, too. Jordana Bell et al., "Oral evidence to UK House of Lords 'Aging: Science, Technology and Healthy Living' Inquiry" (INQ0029) (Science and Technology Committee [House of Lords], 2019) ageless.link/9bajn3.

This paper is a nice review demonstrating how age-related changes don't map neatly onto a single disease, by examining the relative contributions of the many processes which ultimately make cancer an age-related disease: Ezio Laconi, Fabio Marongiu and James DeGregori, "Cancer as a disease of old age: Changing mutational and microenvironmental landscapes," **Br. J. Cancer** 122, 943–52 (2020). DOI: 10.1038/s41416-019-0721-1 ageless.link/c4smzx.

104 **all animals have a fixed number of heartbeats:** H. J. Levine, "Rest heart rate and life expectancy,"

J. Am. Coll. Cardiol. 30, 1104–6 (1997). DOI: 10.1016/s0735-1097(97)00246-5 ageless .link/q34kh7.

The Public Science Lab at NC State University is collating a database of heartbeat and lifespan data which currently includes over 300 animals: The Heart Project, The Public Science Lab, NC State University ageless.link/degeqy.

See also this excellent video on the topic: Rohin Francis, "Why do so many living things get the same number of heartbeats?" (MedLife Crisis, YouTube, 2018) ageless.link/prbvyx.

105 **a higher resting heart rate . . . increased risk of death:** D. Aune et al., "Resting heart rate and the risk of cardiovascular disease, total cancer, and all-cause mortality—a systematic review and dose-response meta-analysis of prospective studies," **Nutr. Metab. Cardiovasc. Dis.** 27, 504–17 (2017). DOI: 10.1016/j.numecd.2017.04.004 ageless.link/ eb3fr9.

105 **it's not clear how . . . a treatment:** Though ultimately it concludes that the best approach to lower heart rate at present is diet and exercise, this paper makes an interesting case for thinking about reducing it medically: Gus Q. Zhang and Weiguo Zhang, "Heart rate, lifespan, and mortality risk," **Aging Res. Rev.** 8, 52–60 (2009). DOI: 10.1016/ j.arr.2008.10.001 ageless.link/hqti9f.

107 **several attempts to systematically classify theories of aging:** One of the first attempts to draw together the large body of disparate knowledge

about aging was written by Alex Comfort in the 1950s. (Comfort was something of a polymath— you may be more familiar with his 1972 book, **The Joy of Sex.**) Alex Comfort, **Aging, the Biology of Senescence** (Routledge & Kegan Paul, 1956) ageless.link/jopnzx.

Another well-known attempt in 1990 reviewed the evolutionary and mechanistic theories of aging in the hope of unifying the subject: Z. A. Medvedev, "An attempt at a rational classification of theories of aging," **Biol. Rev. Camb. Philos. Soc.** 65, 375–98 (1990). DOI: 10.1111/j.1469-185x.1990.tb01428.x ageless.link/ mbs7ot.

107 **"Strategies for Engineered Negligible Senescence":** The original formulation of SENS was published in 2002 as Aubrey D. N. J. de Grey et al., "Time to talk SENS: Critiquing the immutability of human aging," **Ann. N. Y. Acad. Sci.** 959, 452–62 (2002) ageless.link/boetg3.

However, the classification has developed since then, and more recent versions can be found in Ben Zealley and Aubrey D. N. J. de Grey, "Strategies for engineered negligible senescence," **Gerontology** 59, 183–9 (2013). DOI: 10.1159/000342197 ageless.link/ugcyxw. It's also available on the website of de Grey's SENS Research Foundation: Intro to SENS research (SENS Research Foundation) ageless.link/owtoc3 and in his 2008 book: Aubrey de Grey and Michael Rae, **Ending Aging: The Rejuvenation Breakthroughs That Could Reverse Human Aging in Our**

Lifetime (St. Martin's Griffin, 2008) ageless.link/yvitd6.

108 **"The Hallmarks of Aging":** López-Otín et al., 2013 ageless.link/m3gh76.

1. Trouble in the double helix:
DNA damage and mutations

A good review of the evidence for DNA damage and mutations mattering in aging can be found in: Alex A. Freitas and João Pedro de Magalhães, "A review and appraisal of the DNA damage theory of aging," **Mutat. Res.** 728, 12–22 (2011). DOI: 10.1016/j.mrrev.2011.05.001 ageless.link/epodzw.

111 **100,000 assaults on its genetic code every day:** George A. Garinis et al., "DNA damage and aging: New-age ideas for an age-old problem," **Nat. Cell Biol.** 10, 1241–7 (2008). DOI: 10.1038/ncb1108-1241 ageless.link/xp9rgi.

111 **a couple of light-years of DNA:** Data on cell turnover are hard to find and often quite inconsistent because it's an incredibly hard phenomenon to observe experimentally. Luckily for this calculation, by far the majority of DNA replication is due to the huge number of blood cells our bodies produce. You can see the full calculation at ageless.link/969hvc.

112 **people who are successfully treated for cancer in youth:** Rhys Anderson, Gavin D. Richardson and João F. Passos, "Mechanisms driving the aging heart," **Exp. Gerontol.** 109, 5–15 (2018).

DOI: 10.1016/j.exger.2017.10.015 ageless.link/
buov7p.

113 **reduction in life expectancy of about a decade:**
Jennifer M. Yeh et al., "Life expectancy of adult
survivors of childhood cancer over 3 decades,"
JAMA Oncol 6, 350–7 (2020). DOI: 10.1001/
jamaoncol.2019.5582 ageless.link/pouzkf.

2. Trimmed telomeres

116 **Telomeres are also implicated in . . . hair
turning gray:** Mariela Jaskelioff et al., "Telomerase
reactivation reverses tissue degeneration in aged
telomerase-deficient mice," **Nature** 469, 102–6
(2011). DOI: 10.1038/nature09603 ageless.link/
gt7m46.

116 **A study looking at same-sex twins:** Masayuki
Kimura et al., "Telomere length and mortality: A study of leukocytes in elderly Danish
twins," **Am. J. Epidemiol.** 167, 799–806 (2008).
DOI: 10.1093/aje/kwm380 ageless.link/ypcht6.

116 **The largest collection of telomere length data:**
Line Rode, Børge G. Nordestgaard and Stig E.
Bojesen, "Peripheral blood leukocyte telomere
length and mortality among 64,637 individuals
from the general population," **J. Natl. Cancer
Inst.** 107, djv074 (2015). DOI: 10.1093/jnci/
djv074 ageless.link/qkyhcb.

117 **damaged telomeres can signal to a cell that
it's time:** Stella Victorelli and João F. Passos,
"Telomeres and cell senescence—size matters not,"

EBioMedicine 21, 14–20 (2017). DOI: 10.1016/ j.ebiom.2017.03.027 ageless.link/hyrddd.

3. Protein problems:
autophagy, amyloids and adducts

118 **An individual protein molecule . . . will typically last a few days:** Brandon H. Toyama et al., "Identification of long-lived proteins reveals exceptional stability of essential cellular structures," **Cell** 154, 971–82 (2013). DOI: 10.1016/ j.cell.2013.07.037 ageless.link/e96gxu.

119 **[Autophagy's] importance . . . a Nobel Prize in 2016:** The Nobel Prize in Physiology or Medicine 2016: Yoshinori Ohsumi (The Nobel Prize, 2016) ageless.link/x3hxuq.

119 **Reducing or entirely disabling . . . can accelerate aging:** These are both good reviews of the relationship between autophagy, aging and dietary restriction: Andrew M. Leidal, Beth Levine and Jayanta Debnath, "Autophagy and the cell biology of age-related disease," **Nat. Cell Biol.** 20, 1338–48 (2018). DOI: 10.1038/s41556-018 -0235-8 ageless.link/iqycep.

David C. Rubinsztein, Guillermo Mariño and Guido Kroemer, "Autophagy and aging," **Cell** 146, 682–95 (2011). DOI: 10.1016/j.cell.2011.07.030 ageless.link/h3e9va.

119 **age-related diseases . . . with autophagy:** Didac Carmona-Gutierrez et al., "The crucial impact of lysosomes in aging and longevity," **Aging Res. Rev.**

32, 2–12 (2016). DOI: 10.1016/j.arr.2016.04.009 ageless.link/nfc3fm.

120 **Impaired autophagy . . . and heart problems:** Leidal, Levine, and Debnath, 2018 ageless.link/iqycep.

121 **there are now dozens . . . to be implicated:** Tuomas P. J. Knowles, Michele Vendruscolo and Christopher M. Dobson, "The amyloid state and its association with protein misfolding diseases," **Nat. Rev. Mol. Cell Biol.** 15, 384–96 (2014). DOI: 10.1038/nrm3810 ageless.link/qbo7fa.

123 **The Maillard reaction is behind the crust:** Andy Extance, "The marvellous Maillard reaction," **Chemistry World** (2018) ageless.link/pygx4v.

4. Epigenetic alterations

125 **hundreds of different types of cells:** "Cell types" are hotly contested in biology, and assigning a precise number doesn't really make sense. Cells exist on a spectrum rather than in neat pigeonholes— but they're clustered together into families which it's probably worth calling cell types if only to preserve our own sanity when trying to discuss things. This article gathers together several perspectives on this debate: Hans Clevers et al., "What is your conceptual definition of 'cell type' in the context of a mature organism?," **Cell Syst.** 4, 255–9 (2017). DOI: 10.1016/j.cels.2017.03.006 ageless.link/cvj3ba.

126 **The 8,000 samples [Horvath] used in his first paper:** Steve Horvath, "DNA methylation age of

human tissues and cell types," **Genome Biology** 14, R115 (2013). DOI: 10.1186/gb-2013-14-10-r115 ageless.link/gkjacc.

127 **he later told a reporter that he had trouble believing:** This is an accessible account of Horvath's work: W. Wayt Gibbs, "Biomarkers and aging: The clock-watcher," **Nature** 508, 168–70 (2014). DOI: 10.1038/508168a ageless .link/eginsd.

128 **people with an epigenetic age . . . die sooner:** See, for example, Brian H. Chen et al., "DNA methylation-based measures of biological age: Meta-analysis predicting time to death," **Aging** 8, 1844–65 (2016). DOI: 10.18632/ aging.101020 ageless.link/gpji9v.

5. Accumulation of senescent cells

129 **discovered in 1961, by . . . Leonard Hayflick:** J. W. Shay and W. E. Wright, "Hayflick, his limit, and cellular aging," **Nat. Rev. Mol. Cell Biol.** 1, 72–6 (2000). DOI: 10.1038/35036093 ageless .link/dswmot.

129 **overturned half a century of dogma:** The incredible tale of how this dogma was perpetuated by Nobel Prize winner Alexis Carrel and his claims of immortal chicken cells is told in John Rasko and Carl Power, "What pushes scientists to lie? The disturbing but familiar story of Haruko Obokata," **Guardian** (18 February 2015) ageless.link/mbaxre.

131 **just 500,000 senescent cells . . . physical impairment:** Ming Xu et al., "Senolytics improve

physical function and increase lifespan in old age," **Nat. Med.** 24, 1246–56 (2018). DOI: 10.1038/s41591-018-0092-9 ageless.link/kxawt4.

6. Power struggle: malfunctioning mitochondria

Nick Lane's book on mitochondria is a great general introduction to these weird cellular organelles: Nick Lane, **Power, Sex, Suicide: Mitochondria and the Meaning of Life** (Oxford University Press, 2006) ageless.link/6ox4kh.

133 **"fusion" and "fission" events:** Iain Scott and Richard J. Youle, "Mitochondrial fission and fusion," **Essays Biochem.** 47, 85–98 (2010). DOI: 10.1042/bse0470085 ageless.link/k69ons.

134 **There tend to be fewer of them . . . less energy:** Milena Pinto and Carlos T. Moraes, "Mechanisms linking mtDNA damage and aging," **Free Radic. Biol. Med.** 85, 250–58 (2015). DOI: 10.1016/j.freeradbiomed.2015.05.005 ageless.link/wiraa7.

134 **This reduction . . . illness and death:** Stephen Frenk and Jonathan Houseley, "Gene expression hallmarks of cellular aging," **Biogerontology** 19, 547–66 (2018). DOI: 10.1007/s10522-018-9750-z ageless.link/6iuhtc.

134 **mutations in mitochondrial DNA increase with age:** Anne Hahn and Steven Zuryn, "The cellular mitochondrial genome landscape in disease," **Trends Cell Biol.** 29, 227–40 (2019). DOI: 10.1016/j.tcb.2018.11.004 ageless.link/noxwpf.

134 **[Mitophagy] declines with age:** Alexandra

Moreno-García et al., "An overview of the role of lipofuscin in age-related neurodegeneration," **Front. Neurosci.** 12, 464 (2018). DOI: 10.3389/fnins.2018.00464 ageless.link/he6zcr.

134 **damage to mitochondria . . . leads to the loss of muscle mass and strength:** Axel Kowald and Thomas B. L. Kirkwood, "Resolving the enigma of the clonal expansion of mtDNA deletions," **Genes** 9, 126 (2018). DOI: 10.3390/genes9030126 ageless.link/pbsrfj.

135 **dysfunctional mitochondria show up in . . . Parkinson's and Alzheimer's:** Leidal, Levine and Debnath, 2018 ageless.link/iqycep.

135 **Another experiment bred mice with . . . reduced mitochondrial numbers:** Bhupendra Singh et al., "Reversing wrinkled skin and hair loss in mice by restoring mitochondrial function," **Cell Death Dis.** 9, 735 (2018). DOI: 10.1038/s41419-018-0765-9 ageless.link/39uc9a.

136 **the simple idea . . . is an oversimplification:** A slightly advanced but readable review of the overturning of the free radical theory of aging can be found in: David Gems and Linda Partridge, "Genetics of longevity in model organisms: Debates and paradigm shifts," **Annu. Rev. Physiol.** 75, 621–44 (2013). DOI: 10.1146/annurev-physiol-030212-183712 ageless.link/r9g6fx.

7. Signal failure

138 **This process . . . dubbed "inflammaging":** An accessible review of inflammaging can be found

in: Claudio Franceschi and Judith Campisi, "Chronic inflammation (inflammaging) and its potential contribution to age-associated diseases," **J. Gerontol. A Biol. Sci. Med. Sci.** 69 Suppl 1, S4–9 (2014). DOI: 10.1093/gerona/glu057 ageless .link/rzitpw.

8. Gut reaction: changes in the microbiome

Nice reviews of how the microbiome changes with age can be found in:

Thomas W. Buford, "(Dis)Trust your gut: The gut microbiome in age-related inflammation, health, and disease," **Microbiome** 5, 80 (2017). DOI: 10.1186/s40168-017-0296-0 ageless.link/ y49t3u.

Claire Maynard and David Weinkove, "The gut microbiota and aging," in **Biochemistry and Cell Biology of Aging**: Part I **Biomedical Science** (ed. J. Robin Harris and Viktor I. Korolchuk) (Springer Singapore, 2018). DOI: 10.1007/978 -981-13-2835-0_12 ageless.link/fgxork.

Jens Seidel and Dario Riccardo Valenzano, "The role of the gut microbiome during host aging," **F1000Res.** 7, 1086 (2018). DOI: 10.12688/ f1000research.15121.1 ageless.link/gojnhw.

142 **the number of microbial cells in your gut:** A commonly cited statistic is that the cells of our microbiome outnumber our own cells ten to one. This study revises that estimate substantially—but no doubt debate will continue because counting trillions of microbes is not easy!

Ron Sender, Shai Fuchs and Ron Milo, "Are we really vastly outnumbered? Revisiting the ratio of bacterial to host cells in humans," **Cell** 164, 337–40 (2016). DOI: 10.1016/j.cell.2016.01.013 ageless.link/9oeph4.

142 **your intestines can come to be dominated:** Buford, 2017 ageless.link/y49t3u.

143 **we have managed to build "microbial clocks":** Fedor Galkin et al., "Human microbiome aging clocks based on deep learning and tandem of permutation feature importance and accumulated local effects," **bioRxiv** (2018). DOI: 10.1101/507780 ageless.link/3wtnuz.

144 **One study . . . without a microbiome:** Marisa Stebegg et al., "Heterochronic fecal transplantation boosts gut germinal centers in aged mice," **Nat. Commun.** 10, 2443 (2019). DOI: 10.1038/s41467-019-10430-7 ageless.link/srchrr.

9. Cellular exhaustion

146 **increasing their preference for becoming fat cells:** Arantza Infante and Clara I. Rodríguez, "Osteogenesis and aging: Lessons from mesenchymal stem cells," **Stem Cell Res. Ther.** 9, 244 (2018). DOI: 10.1186/s13287-018-0995-x ageless.link/kkbvik.

146 **repeated "compression fractures" . . . shorter with age:** Jerry L. Old and Michelle Calvert, "Vertebral compression fractures in the elderly," **Am. Fam. Physician** 69, 111–16 (2004) ageless.link/u7cuzu.

147 **Olfactory neuron stem cells start to flag:** Lisa Bast et al., "Increasing neural stem cell division asymmetry and quiescence are predicted to contribute to the age-related decline in neurogenesis," **Cell Rep.** 25, 3231–40.e8 (2018). DOI: 10.1016/j.celrep.2018.11.088 ageless.link/9dx7rb.

10. Defective defenses: malfunction of the immune system

A readable general review of the aging of the immune system can be found in: A. Katharina Simon, Georg A. Hollander and Andrew McMichael, "Evolution of the immune system in humans from infancy to old age," **P. Roy. Soc. B: Biol. Sci.** 282, 20143085 (2015). DOI: 10.1098/rspb.2014.3085 ageless.link/b7zdq3.

150 **more than 90 percent . . . over the age of 60:** This was calculated using WHO GBD statistics. You can read about the calculations at ageless.link /x9nrcm.

152 **barely any remains after the age of 60 or so:** Sam Palmer et al., "Thymic involution and rising disease incidence with age," **Proc. Natl. Acad. Sci. U.S.A.** 115, 1883–8 (2018). DOI: 10.1073/pnas.1714478115 ageless.link/sdu6ug.

153 **Memory T cells . . . in the body:** Cornelia M. Weyand and Jörg J. Goronzy, "Aging of the immune system. Mechanisms and therapeutic targets," **Ann. Am. Thorac. Soc.** 13 Suppl 5, S422–S428 (2016). DOI: 10.1513/AnnalsATS .201602-095AW ageless.link/hbxg6g.

153 **T cells specialized . . . a third of our "immune memory":** Paul Klenerman and Annette Oxenius, "T cell responses to cytomegalovirus," **Nat. Rev. Immunol.** 16, 367–77 (2016). DOI: 10.1038/nri.2016.38 ageless.link/f69taa.

5. OUT WITH THE OLD

Killing senescent cells

165 **The first evidence that it does:** Baker et al., 2011 ageless.link/xxobvx.

167 **A study published in 2015:** Yi Zhu et al., "The Achilles' heel of senescent cells: From transcriptome to senolytic drugs," **Aging Cell** 14, 644–58 (2015). DOI: 10.1111/acel.12344 ageless.link/sj9rs3.

168 **But a 2018 study showed that D+Q also has a global effect:** Xu et al., 2018 ageless.link/ijqc4g.

169 **A 2016 study used . . . genetic modification:** Darren J. Baker et al., "Naturally occurring p16Ink4a-positive cells shorten healthy lifespan," **Nature** 530, 184–89 (2016). DOI: 10.1038/nature16932 ageless.link/rkihvv.

169 **The first human clinical trial:** Justice et al., 2019 ageless.link/cx7wkq.

170 **UBX0101 . . . targeting the knees of patients with osteoarthritis:** A study to assess the safety and efficacy of a single dose of UBX0101 in patients with osteoarthritis of the knee (ClinicalTrials.gov identifier: NCT04129944, 2019) ageless.link/d4tcc6.

174 **other approaches to dealing with senescent cells:** This short review describes the range of current therapeutic approaches to cellular senescence: Laura J. Niedernhofer and Paul D. Robbins, "Senotherapeutics for healthy aging," **Nat. Rev. Drug Discov.** 377 (2018). DOI: 10.1038/nrd.2018.44 ageless.link/dkby7o.

　　Another short review looks at senolytics and other anti-aging drugs: Asher Mullard, "Anti-aging pipeline starts to mature," **Nat. Rev. Drug Discov.** 17, 609–12 (2018). DOI: 10.1038/nrd.2018.134 ageless.link/voajt6.

175 **a tube of lotion to speed the healing process:** Scientists have identified a protein called PDGF-AA which seems to be the crucial component of the SASP when it comes to wound healing, at least in mice. We could imagine creating a lotion using this or similar signals to help wounds heal during or shortly after senolytic treatment. Marco Demaria et al., "An essential role for senescent cells in optimal wound healing through secretion of PDGF-AA," **Dev. Cell** 31, 722–33 (2014). DOI: 10.1016/j.devcel.2014.11.012 ageless.link/cwkwyy.

Reinventing recycling: upgrading autophagy

176 **The story of DR mimetics begins:** An account of the remarkable expedition to Rapa Nui can be found in: Amy Tector, "The delightful revolution: Canada's medical expedition to Easter Island, 1964–65," **British Journal of Canadian Studies**

27, 181–94 (2014). DOI: 10.3828/bjcs.2014.12 ageless.link/htyujj.

179 **Scientist Suren Sehgal was dumbfounded:** The story of rapamycin, including Sehgal's part in it, is told in: Bethany Halford, "Rapamycin's secrets unearthed," **Chemical & Engineering News** 94 (2016) ageless.link/7m3abm.

179 **TOR is a nexus in cellular metabolism:** Hannah E. Walters and Lynne S. Cox, "mTORC inhibitors as broad-spectrum therapeutics for age-related diseases," **Int. J. Mol. Sci.** 19 (2018). DOI: 10.3390/ijms19082325 ageless.link/a7dbnk.

179 **rapamycin works . . . already old:** The study reporting that rapamycin extends healthy life in old mice is David E. Harrison et al., "Rapamycin fed late in life extends lifespan in genetically heterogeneous mice," **Nature** 460, 392–5 (2009). DOI: 10.1038/nature08221 ageless.link/af4dtw.

And an analysis of it and its potential implications can be found in Lynne S. Cox, "Live fast, die young: New lessons in mammalian longevity," **Rejuvenation Res.** 12, 283–8 (2009). DOI: 10.1089/rej.2009.0894 ageless.link/r3d3b9.

184 **strengthen their response against COVID-19:** Jamie Metzl and Nir Barzilai, "Drugs that could slow aging may hold promise for protecting the elderly from COVID-19," **Leapsmag** (2020) ageless .link/jvziwf.

184 **There's also spermidine:** This brief, readable article draws together the evidence for spermidine as a DR mimetic drug: Frank Madeo et al.,

"Spermidine delays aging in humans," **Aging** 10, 2209–11 (2018). DOI: 10.18632/aging.101517 ageless.link/qduqki.

184 **Other naturally sourced contenders:** This is a detailed review of current DR mimetic drugs: Frank Madeo et al., "Caloric restriction mimetics against age-associated disease: Targets, mechanisms, and therapeutic potential," **Cell Metab.** 29, 592–610 (2019). DOI: 10.1016/j.cmet.2019.01.018 ageless .link/ovbzfi.

185 **a new mTORC1 inhibitor . . . respiratory infections:** This "Phase 2" study from resTORbio successfully showed these results: Joan B. Mannick et al., "TORC1 inhibition enhances immune function and reduces infections in the elderly," **Sci. Transl. Med.** 10 (2018). DOI: 10.1126/scitranslmed.aaq1564 ageless.link/ ywvxna.

However, a subsequent and subtly different "Phase 3" study was unsuccessful, and analysis is still ongoing: "resTORbio announces that the Phase 3 PROTECTOR 1 trial of RTB101 in clinically symptomatic respiratory illness did not meet the primary endpoint" (resTORbio, Inc., 2019) ageless.link/geknp4.

185 **a new study of RTB101 . . . COVID-19 in nursing home residents:** "resTORbio announces initiation of study to evaluate if antiviral prophylaxis with RTB101 reduces the severity of COVID-19 in nursing home residents" (resTORbio, Inc., 2020) ageless.link/vpzrkn.

186 **Autophagy takes place in . . . lysosomes:** This paper provides a detailed review of lysosomes and lipofuscin in aging: Carmona-Gutierrez et al., 2016 ageless.link/4ksqvf. In particular, the lyosome isn't just a passive recycling plant but sends signals within the cell controlling autophagy and other processes, a fact which isn't discussed due to space constraints.

188 **One of the prime suspects . . . is lipofuscin:** Marcelo M. Nociari, Szilard Kiss and Enrique Rodriguez-Boulan, "Lipofuscin accumulation into and clearance from retinal pigment epithelium lysosomes: Physiopathology and emerging therapeutics," in **Lysosomes: Associated Diseases and Methods to Study Their Function** (ed. Pooja Dhiman Sharma) (InTech, 2017). DOI: 10.5772/intechopen.69304 ageless.link/rrit3y.

188 **lysosomes stuffed to bursting . . . atherosclerotic plaques:** W. Gray Jerome, "Lysosomes, cholesterol and atherosclerosis," **Clin. Lipidol.** 5, 853–65 (2010). DOI: 10.2217/clp.10.70 ageless .link/usc7mq.

189 **lysosomal storage disorders (or LSDs):** For more information about lysosomal storage disorders, see Lysosomal storage disorders (National Organization for Rare Disorders) ageless.link/j4onqe.

190 **North Sea sediment and piles of manure:** Irum Perveen et al., "Studies on degradation of 7-ketocholesterol by environmental bacterial isolates," **Appl. Biochem. Microbiol.** 54, 262–8

(2018). DOI: 10.1134/S0003683818030110 ageless
.link/wctqvb.

190 **cholesterol-crunching   enzymes   in
Mycobacterium tuberculosis:** Brandon M.
D'Arcy et al., "Development of a synthetic
3-ketosteroid δ1-dehydrogenase for the generation of a novel catabolic pathway enabling cholesterol degradation in human cells," **Sci. Rep.** 9,
5969 (2019). DOI: 10.1038/s41598-019-42046-8
ageless.link/f73gvh.

191 **A 2018 paper by . . . Ichor Therapeutics:**
Kelsey J. Moody et al., "Recombinant manganese peroxidase reduces A2E burden in age-
related and Stargardt's macular degeneration
models," **Rejuvenation Res.** 21, 560–71 (2018).
DOI: 10.1089/rej.2018.2146 ageless.link/z7dgq9.

This work is described quite accessibly in a video
from the Life Extension Advocacy Foundation
conference 2019: Kelsey Moody, "Macular degeneration talk at Ending Age-Related Diseases 2019"
(Life Extension Advocacy Foundation, YouTube,
2019) ageless.link/fjekaq.

191 **There's already a drug called Remofuscin:**
F. Yuan et al., "Preclinical results of a new pharmacological therapy approach for Stargardt disease
and dry age-related macular degeneration," **ARVO
2017 E-Abstract** (2017) ageless.link/ojsizw.

192 **get rid of their toxic stockpiles:** Another example is under development by a company called
Underdog Pharmaceuticals, who hope to use
sugars called cyclodextrins to remove oxidized

cholesterol from atherosclerotic plaques: Reason, "An interview with Matthew O'Connor, as Underdog Pharmaceuticals secures seed funding," **Fight Aging!** (2019) ageless.link/c7td7e.

Amyloid

194 **served to undermine this "amyloid hypothesis":** For a particularly critical review of how the amyloid hypothesis came to dominate Alzheimer's research, see Sharon Begley, "How an Alzheimer's 'cabal' thwarted progress toward a cure," **STAT** (2019) ageless.link/tzoitz.

There are also various academic reviews on the subject, e.g., Francesco Panza et al., "A critical appraisal of amyloid-β-targeting therapies for Alzheimer disease," **Nat. Rev. Neurol.** 15, 73–88 (2019). DOI: 10.1038/s41582-018-0116-6 ageless .link/bnu3oy.

197 **similar aggregates . . . in many other diseases:** Knowles, Vendruscolo and Dobson, 2014 ageless .link/y4j4rc.

198 **TTR amyloid may be an underappreciated one:** Yushi Wang et al., "Is vascular amyloidosis intertwined with arterial aging, hypertension and atherosclerosis?," **Front. Genet.** 8, 126 (2017). DOI: 10.3389/fgene.2017.00126 ageless.link/ 9nbz4t.

199 **A study in Finland:** Maarit Tanskanen et al., "Senile systemic amyloidosis affects 25% of the very aged and associates with genetic variation in **alpha2-macroglobulin** and **tau**: A population-

based autopsy study," **Ann. Med.** 40, 232–9 (2008). DOI: 10.1080/07853890701842988 ageless.link/zopekh.

199 **Another in a Spanish hospital:** Esther González-López et al., "Wild-type transthyretin amyloidosis as a cause of heart failure with preserved ejection fraction," **Eur. Heart J.** 36, 2585–94 (2015). DOI: 10.1093/eurheartj/ehv338 ageless.link/mhcfer.

199 **SSA . . . in supercentenarians:** L. Stephen Coles and Robert D. Young, "Supercentenarians and transthyretin amyloidosis: The next frontier of human life extension," **Prev. Med.** 54, S9–S11 (2012). DOI: 10.1016/j.ypmed.2012.03.003 ageless.link/zcbdci.

199 **a drug candidate named PRX004:** Jeffrey N. Higaki et al., "Novel conformation-specific monoclonal antibodies against amyloidogenic forms of transthyretin," **Amyloid** 23, 86–97 (2016). DOI: 10.3109/13506129.2016.1148025 ageless.link/vknf7e.

199 **Also under investigation are catabodies:** Stephanie A. Planque, Richard J. Massey and Sudhir Paul, "Catalytic antibody (catabody) platform for age-associated amyloid disease: From Heisenberg's uncertainty principle to the verge of medical interventions," **Mech. Aging Dev.** 185, 111188 (2020). DOI: 10.1016/j.mad.2019.111188 ageless.link/p3my6b.

200 **GAIM, it was discovered . . . in a bacteriophage:** The story of GAIM and its strange origins is

told in Jon Palfreman, **Brain Storms: The Race to Unlock the Mysteries of Parkinson's Disease** (Scientific American, 2016) ageless.link/mvohpp.

See also Rajaraman Krishnan et al., "A bacteriophage capsid protein provides a general amyloid interaction motif (GAIM) that binds and remodels misfolded protein assemblies," **J. Mol. Biol.** 426, 2500–519 (2014). DOI: 10.1016/j.jmb.2014.04.015 ageless.link/47ffsy.

201 **GAIM has been shown . . . improve their cognitive function:** Jonathan M. Levenson et al., "NPT088 reduces both amyloid-β and tau pathologies in transgenic mice," **Alzheimers. Dement.** 2, 141–55 (2016). DOI: 10.1016/j.trci.2016.06.004 ageless.link/kkt6m9.

6. IN WITH THE NEW

Stem cell therapy

A good general primer on stem cell biology and therapies is Jonathan Slack, **Stem Cells: A Very Short Introduction** (Oxford University Press, 2012) ageless.link/rc4udv.

204 **The term "stem cell" is thrown around by charlatans:** A good resource for evaluating claims about stem cell treatments, as well as general information about stem cells, is "A closer look at stem cells" (International Society for Stem Cell Research) ageless.link/miqgch.

208 **over one million HSC transplants:** Alois Gratwohl et al., "One million haemopoietic

stem-cell transplants: A retrospective obser-vational study," **Lancet Haematol.** 2, e9—00 (2015). DOI: 10.1016/S2352-3026(15)00028-9 ageless.link/qhhjsw.

208 **in the vast majority of cases, HSC donation:** The charity Anthony Nolan has some excellent resources about bone marrow donation, and about blood cancers and other blood disorders more generally.

"Is donating bone marrow painful?" (Anthony Nolan, 2015) ageless.link/qz6una.

209 **there would be no risk of immune rejection either:** In the short term, deriving personalized cells for every patient may be too slow and ex-pensive to make it practical for medical treatment. However, iPSCs will probably make their way into the clinic sooner than fully personalized cells are practical, using iPSC banks containing many dif-ferent people's cells to allow immune compatibil-ity with as many patients as possible. I left this detail out of the main text for simplicity, but this article provides an overview of these plans: Kerry Grens, "Banking on iPSCs," **The Scientist** (2014) ageless.link/vuova4.

211 **a Nobel Prize . . . in 2012:** The Nobel Prize in Physiology or Medicine 2012: Sir John B. Gurdon and Shinya Yamanaka (The Nobel Prize, 2012) ageless.link/9wkqz9.

214 **a tiny, misfolded baby:** Aarathi Prasad, "Teratomas: The tumors that can transform into

'evil twins,'" **Guardian** (27 April 2015) ageless .link/s3dnjp.

215 **Two trials in 2018 . . . patients' eyes:** Lyndon da Cruz et al., "Phase clinical study of an embryonic stem cell-derived retinal pigment epithelium patch in age-related macular degeneration," **Nat. Biotechnol.** 36, 328–37 (2018). DOI: 10.1038/nbt.4114 ageless.link/3srrjw.

Amir H. Kashani et al., "A bioengineered retinal pigment epithelial monolayer for advanced, dry age-related macular degeneration," **Sci. Transl. Med.** 10, eaao4097 (2018). DOI: 10.1126/scitranslmed.aao4097 ageless.link/ayocq6.

215 **The first test in humans . . . stopped for safety reasons:** Ken Garber, "RIKEN suspends first clinical trial involving induced pluripotent stem cells," **Nat. Biotechnol.** 33, 890–9 (2015). DOI: 10.1038/nbt0915-890 ageless.link/iaocp6.

216 **A 2019 study by the U.S. National Eye Institute:** Sharon Begley, "Trial will be first in U.S. of Nobel-winning stem cell technique," **STAT** (2019) ageless .link/pzxvg7.

216 **Stem cell treatments for Parkinson's disease:** The story of using stem cells for Parkinson's is well told in this article, written by two of the Swedish scientists who began this work in the 1980s.

Anders Björklund and Olle Lindvall, "Replacing dopamine neurons in Parkinson's disease: How did it happen?," **J. Parkinsons. Dis.** 7, S21–S31

(2017). DOI: 10.3233/JPD-179002 ageless.link/ hcz3an.

You can read more about stem cells as a treatment for Parkinson's in Palfreman, 2016 ageless .link/h3afof.

Improving immunity

A good general overview of the topics in this section can be found in Richard Aspinall and Wayne A. Mitchell, "The future of aging— pathways to human life extension" (ed. L. Stephen Coles, Gregory M. Fahy and Michael D. West) (Springer, 2010) ageless.link/4tf6gj.

A more technical and more recent review is Janko Nikolich-Žugich, "The twilight of immunity: Emerging concepts in aging of the immune system," **Nat. Immunol.** 19, 10–19 (2018). DOI: 10.1038/s41590-017-0006-x ageless .link/doaepd.

220 **didn't differ from other male singers:** A review of the evidence for sex hormones' involvement in longevity (particularly of men) can be found in David Gems, "Evolution of sexually dimorphic longevity in humans," **Aging** 6, 84–91 (2014). DOI: 10.18632/aging.100640 ageless.link/ b9mxgx.

The castrati study specifically is J. S. Jenkins, "The voice of the castrato," **Lancet** 351, 1877–80 (1998). DOI: 10.1016/s0140-6736(97)10198-2 ageless.link/7pxy9m.

221 **inmates of an institution for the "mentally**

retarded": This study is worth reading just to see how appallingly people with learning disabilities were treated, even in the academic literature, as recently as 1969. One table groups "mental deficiency" into "normal," "borderline," "moronic," "imbecilic" and "idiotic."

J. B. Hamilton and G. E. Mestler, "Mortality and survival: Comparison of eunuchs with intact men and women in a mentally retarded population," **J. Gerontol.** 24, 395–411 (1969). DOI: 10.1093/geronj/24.4.395 ageless.link/i7q6qk.

221 **an analysis of the eunuchs in the Korean Joseon dynasty:** Kyung-Jin Min, Cheol-Koo Lee and Han-Nam Park, "The lifespan of Korean eunuchs," **Curr. Biol.** 22, R792–3 (2012). DOI: 10.1016/j.cub.2012.06.036 ageless.link/7csyw7.

222 **Ki-Won Lee . . . lived under five:** The birth and death dates of the three centenarian eunuchs are as follows: Gyeong-Heon Gi (1670–1771); In-Bo Hong (1735–1835); Ki-Won Lee (1784–1893). (Kyung-Jin Min, personal communication, 2020).

222 **castrating nine-month-old mice:** Tracy S. P. Heng et al., "Impact of sex steroid ablation on viral, tumor and vaccine responses in aged mice," **PLoS One** 7, e42677 (2012). DOI: 10.1371/journal.pone.0042677 ageless.link/rdzawt.

223 **a small human trial . . . Intervene Immune:** Gregory M. Fahy et al., "Reversal of epigenetic aging and immunosenescent trends in humans," **Aging Cell** 18, e13028 (2019). DOI: 10.1111/acel.13028 ageless.link/ebi7qv.

224 **<u>FOXN1</u> seems to be capable . . . driving thymic regeneration:** "Engage reverse gear," *The Economist* (8 April 2014) ageless.link/n946he.

Nicholas Bredenkamp, Craig S. Nowell and C. Clare Blackburn, "Regeneration of the aged thymus by a single transcription factor," **Development** 141, 1627–37 (2014). DOI: 10.1242/dev.103614 ageless.link/gmzmrm.

225 **"thymus organoids" . . . grown in the lab:** Asako Tajima et al., "Restoration of thymus function with bioengineered thymus organoids," **Curr. Stem Cell Rep.** 2, 128–39 (2016). DOI: 10.1007/ s40778-016-0040-x ageless.link/kqdsmo.

226 **T cells need functioning lymph nodes:** Heather L. Thompson et al., "Lymph nodes as barriers to T-cell rejuvenation in aging mice and nonhuman primates," **Aging Cell** 18, e12865 (2019). DOI: 10.1111/acel.12865 ageless.link/bckcdq.

228 **CMV antibodies . . . 40 percent more likely to die:** Eric T. Roberts et al., "Cytomegalovirus antibody levels, inflammation, and mortality among elderly Latinos over 9 years of follow-up," **Am. J. Epidemiol.** 172, 363–71 (2010). DOI: 10.1093/ aje/kwq177 ageless.link/7qdqtt.

229 **human and economic case for CMV vaccine research:** Ann M. Arvin et al., "Vaccine development to prevent cytomegalovirus disease: Report from the national vaccine advisory committee," **Clin. Infect. Dis.** 39, 233–9 (2004). DOI: 10.1086/421999 ageless.link/7eaydz.

231 **if one identical twin develops MS:** Alastair

Compston and Alasdair Coles, "Multiple sclerosis," **Lancet** 372, 1502–17 (2008). DOI: 10.1016/S0140-6736(08)61620-7 ageless.link/hku6nx.

231 **HSC transplants have a higher success rate:** Paolo A. Muraro et al., "Autologous hematopoietic stem cell transplantation for treatment of multiple sclerosis," **Nat. Rev. Neurol.** 13, 391–405 (2017). DOI: 10.1038/nrneurol.2017.81 ageless .link/w3pd3x.

231 **Immune reboots . . . inflammatory bowel disease and lupus:** John A. Snowden, "Rebooting autoimmunity with autologous HSCT," **Blood** 127, 8–10 (2016). DOI: 10.1182/blood-2015-11-678607 ageless.link/viww9d.

232 **neither patient has detectable levels of the virus:** Ravindra Kumar Gupta et al., "Evidence for HIV-1 cure after **CCR5Δ32/Δ32** allogeneic haemopoietic stem-cell transplantation 30 months post ana—lytical treatment interruption: A case report," **Lancet HIV** 7, e340–e347 (2020). DOI: 10.1016/S2352-3018(20)30069-2 ageless.link/6kaq6f.

232 **Scientists in Texas transplanted HSCs:** Michael J. Guderyon et al., "Mobilization-based transplantation of young-donor hematopoietic stem cells extends lifespan in mice," **Aging Cell** 19, e13110(2020). DOI: 10.1111/acel.13110 ageless .link/nvjnw7.

232 **Another group in Los Angeles destroyed:** Melanie M. Das et al., "Young bone marrow transplantation preserves learning and memory in old mice," **Commun. Biol.** 2, 73 (2019).

DOI: 10.1038/s42003-019-0298-5 ageless.link/ 7zqmf4.

233 **mortality in MS patients:** Muraro et al., 2017 ageless.link/w3pd3x.

234 **work is ongoing to make HSC transplants safer:** Akanksha Chhabra et al., "Hematopoietic stem cell transplantation in immunocompetent hosts without radiation or chemotherapy," **Sci. Transl. Med.** 8, 351ra105 (2016). DOI: 10.1126/ scitranslmed.aae0501 ageless.link/k6g7qu.

Modifying the microbiome

Reviews of changes in the microbiome with aging and prospects for treatment can be found in Written evidence to UK House of Lords "Aging: Science, Technology and Healthy Living" Inquiry (INQ0029) (Society for Applied Microbiology, 2019) ageless.link/6r9jp7.

Maynard and Weinkove 2018 ageless.link/ eitcnv.

Buford, 2017 ageless.link/o44mop.

236 **A probiotic cocktail . . . SLAB51:** Laura Bonfili et al., "Gut microbiota manipulation through probiotics oral administration restores glucose homeostasis in a mouse model of Alzheimer's disease," **Neurobiol. Aging** 87, 35–43 (2019). DOI: 10.1016/j.neurobiolaging.2019.11.004 ageless.link/jjwfum.

236 **Probiotics, prebiotics and synbiotics:** Elmira Akbari et al., "Effect of probiotic supplementation on cognitive function and metabolic status in

Alzheimer's disease: A randomized, double-blind and controlled trial," **Front. Aging Neurosci.** 8, 256 (2016). DOI: 10.3389/fnagi.2016.00256 ageless.link/vmbxu3.

237 **the turquoise killifish . . . shortest-lived vertebrates:** Jason Daley, "Meet the fish that grows up in just 14 days," **Smithsonian Magazine** (8 August 2018) ageless.link/knpsfy.

Itamar Harel et al., "A platform for rapid exploration of aging and diseases in a naturally short-lived vertebrate," **Cell** 160, 1013–26 (2015). DOI: 10.1016/j.cell.2015.01.038 ageless.link/3brwe3.

237 **Researchers used the killifish:** Patrick Smith et al., "Regulation of life span by the gut microbiota in the short-lived African turquoise killifish," **Elife** 6 (2017). DOI: 10.7554/eLife.27014 ageless.link/iekcdn.

238 **microbiome transplant . . . able to extend their lifespan:** Clea Bárcena et al., "Healthspan and lifespan extension by fecal microbiota transplantation into progeroid mice," **Nat. Med.** (2019). DOI: 10.1038/s41591-019-0504-5 ageless.link/fx9gzp.

239 **an audaciously exhaustive experiment:** Bing Han et al., "Microbial genetic composition tunes host longevity," **Cell** 169, 1249–1262.e13 (2017). DOI: 10.1016/j.cell.2017.05.036 ageless.link/zxtwy4.

There's also a great feature article about this work: Ed Yong, "A tiny tweak to gut bacteria can extend an animal's life (. . . at least in worms. Would it work in humans?)," **The Atlantic** (15 June 2017) ageless.link/zb3wgi.

Keeping protein pristine

A detailed review of the ways that proteins outside of cells go wrong with aging can be found in Helen L. Birch, "Extracellular matrix and aging," in **Biochemistry and Cell Biology of Aging**: Part I, **Biomedical Science** (ed. J. Robin Harris and Viktor I. Korolchuk) (Springer Singapore, 2018). DOI: 10.1007/978-981-13-2835-0_7 ageless.link/sxmcr9.

241 **held together by crosslinks:** Collagen and its crosslinks are actually even more incredible, because some of them break and re-form when the collagen is stretched, meaning that huge numbers of tiny, reversible chemical reactions are another reason our collagen has exactly the right elasticity.

  Melanie Stammers et al., "Mechanical stretching changes cross-linking and glycation levels in the collagen of mouse tail tendon," **J. Biol. Chem.** 295, 10572–10580 (2020). DOI: 10.1074/jbc .RA119.012067 ageless.link/cz9gtr.

242 **Highly reactive chemicals . . . widespread disruption:** David M. Hudson et al., "Glycation of type I collagen selectively targets the same helical domain lysine sites as lysyl oxidase-mediated cross-linking," **J. Biol. Chem.** 293, 15620–27 (2018). DOI: 10.1074/jbc.RA118.004829 ageless .link/saeoez.

242 **AGE, which is permanent:** David R. Sell and Vincent M. Monnier, "Molecular basis of arterial stiffening: Role of glycation—a mini-review," **Gerontology** 58, 227–37 (2012). DOI: 10.1159/000334668 ageless.link/7qczho.

243 **feedback loops that make matters worse:** Megan A. Cole et al., "Extracellular matrix regulation of fibroblast function: Redefining our perspective on skin aging," **J. Cell Commun. Signal.** 12, 35–43 (2018). DOI: 10.1007/s12079-018 -0459-1 ageless.link/fwyar4.

245 **calling this received wisdom into question:** Stammers et al., **J. Biol. Chem.** (2020). ageless .link/vmvrow.

Sneha Bansode et al., "Glycation changes molecular organization and charge distribution in type I collagen fibrils," **Sci. Rep.** 10, 3397 (2020). DOI: 10.1038/s41598-020-60250-9 ageless.link/ udr6zg.

245 **Scientists are working on "AGE-breaker" drugs:** Nam Y. Kim et al., "Biocatalytic reversal of advanced glycation end product modification," **Chembiochem** 20, 2402–10 (2019). DOI: 10.1002/cbic.201900158 ageless.link/ 36buaw.

Drug development of AGE-breakers substantially predates this, however: one called alagebrium showed promise in rats, dogs and even monkeys, but never quite worked in people, for reasons that are still unclear. (Confusingly, the leading theory is that it never was an AGE-breaker in the first place, but its success was based on other effects.) A good review can be found in Sell and Monnier, 2012 ageless.link/7qczho.

247 **neutrophils rampage:** Elizabeth Sapey et al., "Phosphoinositide 3-kinase inhibition restores neutrophil accuracy in the elderly: Toward targeted treatments for immunosenescence," **Blood** 123, 239–48 (2014). DOI: 10.1182/blood-2013 -08-519520 ageless.link/h7h4zx.

An account of this work is given in Sue Armstrong, **Borrowed Time: The Science of How and Why We Age,** Chapter 9 (Bloomsbury Sigma, 2019) ageless.link/zz7mje.

7. RUNNING REPAIRS

Telomere extensions

This section discusses the work of María Blasco and her group. She gives an excellent overview in this talk: Maria A. Blasco, "Telomeres talk at Ending Age-Related Diseases 2019" (Life Extension Advocacy Foundation, YouTube, 2019) ageless .link/74nqov.

For a more advanced review of telomeres and telomerase therapies, see Paula Martínez and Maria A. Blasco, "Telomere-driven diseases and telomere-targeting therapies," **J. Cell Biol.** 216, 875–87 (2017). DOI: 10.1083/jcb.201610111 ageless.link/bimqri.

251 **[Elizabeth Blackburn and Carol Greider]** . . . **with Jack Szostak:** The Nobel Prize in Physiology

or Medicine 2009: Elizabeth H. Blackburn, Carol W. Greider and Jack W. Szostak (The Nobel Prize, 2009) ageless.link/hawwqj.

251 **ironically, using cells belonging to Leonard Hayflick:** An interview with Hayflick about donating his skin to Geron is available on YouTube. The relevant section starts 37 minutes into the video.

"Back to immortality: Episode 3, Alexis Carrel, Hayflick, telomeres, and cellular aging" (Michael D. West, YouTube, 2017) ageless.link/kpmgcn.

253 **Scientists added extra copies of the telomerase gene:** Steven E. Artandi et al., "Constitutive telomerase expression promotes mammary carcinomas in aging mice," **Proc. Natl. Acad. Sci. U.S.A.** 99, 8191–6 (2002). DOI: 10.1073/pnas.112515399 ageless.link/jju6vq.

253 **a lack of the enzyme suppressed tumor growth:** E. González-Suárez et al., "Telomerase-deficient mice with short telomeres are resistant to skin tumorigenesis," **Nat. Genet.** 26, 114–17 (2000). DOI: 10.1038/79089 ageless.link/cky6h7.

254 **Different species . . . telomerase tightrope:** The interspecies differences in telomere dynamics are fascinating, but there wasn't space to go into huge detail about them. One interesting theory is that it's not the absolute length of telomere that is important, but the interplay between that length and the rate at which they shorten. This paper performs a cross-species analysis supporting that

conclusion: Kurt Whittemore et al., "Telomere shortening rate predicts species life span," **Proc. Natl. Acad. Sci. U.S.A.** 116 (30) 15122–15127 (2019). DOI: 10.1073/pnas.1902452116 ageless .link/gm3fxu.

254 **rare genetic disease . . . dyskeratosis congenita:** M. Soledad Fernández García and Julie Teruya-Feldstein, "The diagnosis and treatment of dyskeratosis congenita: A review," **J. Blood Med.** 5, 157–67 (2014). DOI: 10.2147/JBM.S47437 ageless.link/66ttiu.

255 **At the opposite extreme, a family was found in Germany:** Susanne Horn et al., "TERT promoter mutations in familial and sporadic melanoma," **Science** 339, 959–61 (2013). DOI: 10.1126/ science.1230062 ageless.link/icwi7k.

256 **natural variation . . . doesn't much matter overall:** Telomeres Mendelian Randomization Collaboration et al., "Association between telomere length and risk of cancer and non-neoplastic diseases: A Mendelian randomization study," **JAMA Oncol.** 3, 636–51 (2017). DOI: 10 .1001/jamaoncol.2016.5945 ageless.link/jvvudx.

257 **In 2008 . . . extend lifespan in mice:** Antonia Tomás-Loba et al., "Telomerase reverse transcriptase delays aging in cancer-resistant mice," **Cell** 135, 609–22 (2008). DOI: 10.1016/ j.cell.2008.09.034 ageless.link/36fh7o.

257 **A follow-up . . . in adult mice:** Bruno Bernardes de Jesus et al., "Telomerase gene therapy in

adult and old mice delays aging and increases longevity without increasing cancer," **EMBO Mol. Med.** 4, 691–704 (2012). DOI: 10.1002/emmm.201200245 ageless.link/cq3dcf.

258 **the same viral gene therapy . . . cancer susceptibility:** Miguel A. Muñoz-Lorente et al., "AAV9-mediated telomerase activation does not accelerate tumorigenesis in the context of oncogenic K-Ras-induced lung cancer," **PLoS Genet.** 14, e1007562 (2018). DOI: 10.1371/journal.pgen.1007562 ageless.link/ft9h9w.

258 **very long telomeres, but completely normal telomerase:** Miguel A. Muñoz-Lorente, Alba C. Cano-Martin and Maria A. Blasco, "Mice with hyper-long telomeres show less metabolic aging and longer lifespans," **Nat. Commun.** 10, 4723 (2019). DOI: 10.1038/s41467-019-12664-x ageless.link/n7rx99.

259 **Experiments . . . can reverse IPF:** Juan Manuel Povedano et al., "Therapeutic effects of telomerase in mice with pulmonary fibrosis induced by damage to the lungs and short telomeres," **Elife** 7, e31299 (2018). DOI: 10.7554/eLife.31299 ageless.link/syg3of.

260 **The most studied is TA-65:** Martínez and Blasco, 2017 ageless.link/bimqri.

Can young blood teach old cells new tricks?

This article is an accessible and in-depth treatment of modern heterochronic parabiosis research:

Megan Scudellari, "Aging research: Blood to blood," **Nature** 517, 426–9 (2015). DOI: 10.1038/517426a ageless.link/nyionc.

An overview of the history of parabiosis can be found in Michael J. Conboy, Irina M. Conboy and Thomas A. Rando, "Heterochronic parabiosis: Historical perspective and methodological considerations for studies of aging and longevity," **Aging Cell** 12, 525–30 (2013). DOI: 10.1111/acel.12065 ageless.link/cjhjti.

262 **In 1864, physiologist Paul Bert:** Clive M. McCay et al., "Parabiosis between old and young rats," **Gerontologia** 1, 7–17 (1957) ageless.link/gmtdab.

262 **The tooth decay experiment:** B. B. Kamrin, "Local and systemic cariogenic effects of refined dextrose solution fed to one animal in parabiosis," **J. Dent. Res.** 33, 824–9 (1954). DOI: 10.1177/00 220345540330061001 ageless.link/f6gxif.

263 **in the fifties, performed by Clive McCay:** McCay et al., 1957 ageless.link/gmtdab.

264 **Experiments . . . more robust picture:** Frederic C. Ludwig and Robert M. Elashoff, "Mortality in syngeneic rat parabionts of different chronological age," **Trans. N. Y. Acad. Sci.** 34, 582–7 (1972). DOI: 10.1111/j.2164-0947.1972.tb02712.x ageless .link/igskpz.

265 **results were clear-cut:** Irina M. Conboy et al., "Rejuvenation of aged progenitor cells by exposure to a young systemic environment," **Nature** 433, 760–64 (2005). DOI: 10.1038/nature03260 ageless.link/67itru.

268 **improved growth . . . in the brain:** Lida Katsimpardi et al., "Vascular and neurogenic rejuvenation of the aging mouse brain by young systemic factors," **Science** 344, 630–34 (2014). DOI: 10.1126/science.1251141 ageless.link/eb6qyi.

268 **better spinal cord regeneration:** Julia M. Ruckh et al., "Rejuvenation of regeneration in the aging central nervous system," **Cell Stem Cell** 10, 96–103 (2012). DOI: 10.1016/j.stem.2011.11.019 ageless.link/7x7w6k.

268 **and can have an aged, oversized heart shrunk:** Francesco S. Loffredo et al., "Growth differentiation factor 11 is a circulating factor that reverses age-related cardiac hypertrophy," **Cell** 153, 828–39 (2013). DOI: 10.1016/j.cell.2013.04.015 ageless .link/9qbpim.

269 **one in South Korea hoping to use young plasma:** Myung Ryool Park, "Clinical trial to evaluate the potential efficacy and safety of human umbilical cord blood and plasma" (ClinicalTrials .gov identifier NCT02418013, 2015) ageless.link/ rp7apo.

269 **a U.S. trial gave transfusions . . . to Alzheimer's patients:** Sharon J. Sha et al., "Safety, tolerability, and feasibility of young plasma infusion in the plasma for Alzheimer symptom amelioration study: A randomized clinical trial," **JAMA Neurol.** 76, 35–40 (2019). DOI: 10.1001/ jamaneurol.2018.3288 ageless.link/d33ozp.

269 **One colorful outfit called Ambrosia:** Zoë Corbyn, "Could 'young' blood stop us getting

old?," **Guardian** (2 February 2020) ageless.link/
mv4fhr.

269 **Rumors that Peter Thiel:** Jeff Bercovici, "Peter
Thiel is very, very interested in young people's
blood," **Inc.** (2016) ageless.link/wmadgf.

270 **[young plasma] didn't make them live any
longer:** Dmytro Shytikov et al., "Aged mice repeatedly injected with plasma from young mice:
A survival study," **Biores. Open Access** 3, 226–32
(2014). DOI: 10.1089/biores.2014.0043 ageless
.link/4vrkko.

270 **young plasma . . . improve liver function in old
mice:** Anding Liu et al., "Young plasma reverses
age-dependent alterations in hepatic function
through the restoration of autophagy," **Aging Cell**
17 (2018). DOI: 10.1111/acel.12708 ageless.link/
sbjw6a.

270 **connecting pairs of rodents to a tiny pumping
device:** Justin Rebo et al., "A single heterochronic
blood exchange reveals rapid inhibition of multiple tissues by old blood," **Nat. Commun.** 7, 13363
(2016). DOI: 10.1038/ncomms13363 ageless.link/
kcavhd.

272 **One age-related miscreant identified is . . .
TGF-beta:** Hanadie Yousef et al., "Systemic at-
tenuation of the TGF-β pathway by a single drug
simultaneously rejuvenates hippocampal neurogenesis and myogenesis in the same old mammal,"
Oncotarget 6, 11959–78 (2015). DOI: 10.18632/
oncotarget.3851 ageless.link/aonk34.

272 **oxytocin . . . is a potential beneficial factor:**

Christian Elabd et al., "Oxytocin is an age-specific circulating hormone that is necessary for muscle maintenance and regeneration," **Nat. Commun.** 5, 4082 (2014). DOI: 10.1038/ncomms5082 ageless.link/cdmifq.

272 **A protein called GDF11:** Manisha Sinha et al., "Restoring systemic GDF11 levels reverses age-related dysfunction in mouse skeletal muscle," **Science** 344, 649–52 (2014). DOI: 10.1126/science.1251152 ageless.link/fr9etf.

273 **dialing down the activity of TGF-beta:** Yousef et al., 2015 ageless.link/aonk34.

273 **the drug and extra oxytocin:** Melod Mehdipour et al., "Rejuvenation of brain, liver and muscle by simultaneous pharmacological modulation of two signaling determinants, that change in opposite directions with age," **Aging** 11, 5628–45 (2019). DOI: 10.18632/aging.102148 ageless.link/n9nfvg.

274 **One study . . . the hypothalamus:** Yalin Zhang et al., "Hypothalamic stem cells control aging speed partly through exosomal miRNAs," **Nature** 548, 52–57 (2017). DOI: 10.1038/nature23282 ageless.link/bu3kdh.

Powering up mitochondria

This review provides a good overview of mitochondria in aging, concentrating mainly on mitochondrial mutations as a root cause: James B. Stewart and Patrick F. Chinnery, "The dynamics of mitochondrial DNA heteroplasmy: Implications for human health and disease," **Nat. Rev. Genet.** 16, 530–42 (2015). DOI: 10.1038/nrg3966 ageless.link/epiywo.

278 **A Cochrane systematic review:** Goran Bjelakovic et al., "Antioxidant supplements for prevention of mortality in healthy participants and patients with various diseases," **Cochrane Database Syst. Rev.** CD007176 (2012). DOI: 10.1002/14651858. CD007176.pub2 ageless.link/guchwk.

280 **they lived 20 percent longer than regular mice:** Samuel E. Schriner et al., "Extension of murine life span by overexpression of catalase targeted to mitochondria," **Science** 308, 1909–11 (2005). DOI: 10.1126/science.1106653 ageless.link/nwcpsg.

280 **mitochondrially . . . risk of cancer:** Xuang Ge et al., "Mitochondrial catalase suppresses naturally occurring lung cancer in old mice," **Pathobiol. Aging Age Relat. Dis.** 5, 28776 (2015). DOI: 10.3402/pba.v5.28776 ageless.link/fqtqeq.

280 **slow . . . age-related heart problems:** Dao-Fu Dai et al., "Overexpression of catalase targeted to mitochondria attenuates murine cardiac aging," **Circulation** 119, 2789–97 (2009). DOI: 10.1161/CIRCULATIONAHA.108.822403 ageless.link/voxv4s.

280 **reduce the production of amyloid-beta:** Peizhong Mao et al., "Mitochondria-targeted catalase reduces abnormal APP processing, amyloid β production and BACE1 in a mouse model of Alzheimer's disease: Implications for

neuroprotection and lifespan extension," **Hum. Mol. Genet.** 21, 2973–90 (2012). DOI: 10.1093/hmg/dds128 ageless.link/divufs.

280 **and improve muscle function:** Alisa Umanskaya et al., "Genetically enhancing mitochondrial antioxidant activity improves muscle function in aging," **Proc. Natl. Acad. Sci. U.S.A.** 111, 15250–55 (2014). DOI: 10.1073/pnas.1412754111 ageless.link/eh3aty.

280 **human clinical trials [of MitoQ] . . . reduce inflammation of the liver:** Edward J. Gane et al., "The mitochondria-targeted anti-oxidant mito-quinone decreases liver damage in a phase II study of hepatitis C patients: Mitoquinone and liver damage," **Liver Int.** 30, 1019–26 (2010). DOI: 10.1111/j.1478-3231.2010.02250.x ageless .link/cshjfw.

280 **improved blood vessel function:** Matthew J. Rossman et al., "Chronic supplementation with a mitochondrial antioxidant (MitoQ) improves vascular function in healthy older adults," **Hypertension** 71, 1056–63 (2018). DOI: 10.1161/HYPERTENSIONAHA.117.10787 ageless.link/cmtudh.

280 **didn't slow the progression of Parkinson's disease:** Huajun Jin et al., "Mitochondria-targeted antioxidants for treatment of Parkinson's disease: preclinical and clinical outcomes," **Biochim. Biophys. Acta** 1842, 1282–94 (2014). DOI: 10.1016/j.bbadis.2013.09.007 ageless.link/qstzg4.

280 **MitoQ . . . decrease the rate at which they shorten:** Victorelli and Passos, 2017 ageless.link/ rb3hdo.

281 **urolithin A . . . extend lifespan in worms, improve . . . mice:** Dongryeol Ryu et al., "Urolithin A induces mitophagy and prolongs lifespan in **C. elegans** and increases muscle function in rodents," **Nat. Med.** 22, 879–88 (2016). DOI: 10.1038/nm.4132 ageless.link/6aknqr.

281 **slow cognitive decline . . . of Alzheimer's:** Zhuo Gong et al., "Urolithin A attenuates memory impairment and neuroinflammation in APP/ PS1 mice," **J. Neuroinflammation** 16, 62 (2019). DOI: 10.1186/s12974-019-1450-3 ageless.link/ a7whwj.

281 **improve mitochondrial function in people:** Pénélope A. Andreux et al., "The mitophagy activator urolithin A is safe and induces a molecular signature of improved mitochondrial and cellular health in humans," **Nature Metabolism** 1, 595–603 (2019). DOI: 10.1038/s42255-019 -0073-4 ageless.link/qvjn9c.

281 **NAD⁺:** Evandro F. Fang et al., "NAD⁺ in aging: molecular mechanisms and translational implications," **Trends Mol. Med.** 23, 899–916 (2017). DOI: 10.1016/j.molmed.2017.08.001 ageless.link/g9fw7e.

283 **ATP8 gene . . . a copy in the nucleus:** D. P. Gearing and P. Nagley, "Yeast mitochondrial ATPase subunit 8, normally a mitochondrial gene product, expressed in vitro and imported back

into the organelle," **EMBO J.** 5, 3651–5 (1986) ageless.link/w6en34.

283 **a therapy . . . LHON:** Yong Zhang et al., "The progress of gene therapy for Leber's optic hereditary neuropathy," **Curr. Gene Ther.** 17, 320–26 (2017). DOI: 10.2174/1566523218666171129204 926 ageless.link/inirfc.

284 **restoring function . . . backup in the nucleus:** Amutha Boominathan et al., "Stable nuclear expression of ATP8 and ATP6 genes rescues a mtDNA complex V null mutant," **Nucleic Acids Res.** 44, 9342–57 (2016). DOI: 10.1093/nar/gkw756 ageless.link/nqcgrj.

284 **getting all 13 mitochondrially encoded genes working:** Caitlin J. Lewis et al., "Codon optimization is an essential parameter for the efficient allotopic expression of mtDNA genes," **Redox Biol.** 30, 101429 (2020). DOI: 10.1016/j.redox.2020.101429 ageless.link/kpmpte.

284 **why hasn't evolution already done it?:** The various reasons evolution might not have already moved the mitochondrial genes to the nucleus are explored in this paper. One idea I've not mentioned is that the proteins retained in mitochondria might be too "hydrophobic"—prone to bending out of shape when coming into contact with water—which would make their construction in and journey through the water-filled cell interior to the mitochondria impossible without destroying their structure. The "local government" metaphor is more properly called the "colocalization

for redox regulation" (CoRR) hypothesis, and is also discussed.

Iain G. Johnston and Ben P. Williams, "Evolutionary inference across eukaryotes identifies specific pressures favoring mitochondrial gene retention," **Cell Syst.** 2, 101–11 (2016). DOI: 10.1016/j.cels.2016.01.013 ageless.link/4i66ik.

285 **dominated by zombie mitochondrial clones:** Kowald and Kirkwood, 2018 ageless.link/s9qfqu.

Repelling the attack of the clones

This is a short, readable review making the case, as this section does, for the importance of clonal expansions in aging: Inigo Martincorena, "Somatic mutation and clonal expansions in human tissues," **Genome Med.** 11, 35 (2019). DOI: 10.1186/s13073-019-0648-4 ageless.link/gg3ix4.

288 **taller people are at greater risk:** Leonard Nunney, "Size matters: Height, cell number and a person's risk of cancer," **Proc. Biol. Sci.** 285 (2018). DOI: 10.1098/rspb.2018.1743 ageless.link/iasikc.

289 **Don't panic, tall people:** Emelie Benyi et al., "Adult height is associated with risk of cancer and mortality in 5.5 million Swedish women and men," **J. Epidemiol. Community Health** 73, 730–36 (2019). DOI: 10.1136/jech-2018-211040 ageless.link/aobtr4.

289 **The elephant genome contains:** Michael Sulak et al., "**TP53** copy number expansion is associated with the evolution of increased body size and an

enhanced DNA damage response in elephants," **Elife** 5, e11994 (2016). DOI: 10.7554/eLife.11994 ageless.link/u4uzsy.

289 **Bowhead whales don't have extra p53:** Michael Keane et al., "Insights into the evolution of longevity from the bowhead whale genome," **Cell Rep.** 10, 112–22 (2015). DOI: 10.1016/j.celrep.2014.12.008 ageless.link/yc3ucj.

292 **One of a handful of exceptions is sun-exposed skin:** Iñigo Martincorena et al., "High burden and pervasive positive selection of somatic mutations in normal human skin," **Science** 348, 880–86 (2015). DOI: 10.1126/science.aaa6806 ageless.link/r33c9h.

293 **the lining of your lungs if you're a smoker:** Kenichi Yoshida et al., "Tobacco smoking and somatic mutations in human bronchial epithelium," **Nature** 578, 266–72 (2020). DOI: 10.1038/s41586-020-1961-1 ageless.link/dyefiz.

294 **specific faults in its genome:** The characteristics I go on to list are a summary of the "hallmarks of cancer"—a famous paper whose approach inspired the hallmarks of aging. The original hallmarks were published in 2000, and this is an update: Douglas Hanahan and Robert A. Weinberg, "Hallmarks of cancer: The next generation," **Cell** 144, 646–74 (2011). DOI: 10.1016/j.cell.2011.02.013 ageless.link/ut79vk.

296 **around half of us are predicted to be diagnosed:** Cancer Research UK maintains fantastic statistical resources on cancer risk, number of

deaths and much more. Their statistics are mainly UK-specific, but numbers are similar across the rich world.

Lifetime risk of cancer (Cancer Research UK, 2015) ageless.link/yqazjf.

297 **2015 study looking at the skin of four people:** Martincorena et al., 2015 ageless.link/r33c9h.

298 **Subsequent work looking at the esophagus:** Iñigo Martincorena et al., "Somatic mutant clones colonize the human esophagus with age," **Science** 362, 911–17 (2018). DOI: 10.1126/science.aau3879 ageless.link/9okjc3.

298 **The protein it encodes:** Technically, **DNMT3A** is responsible for changing DNA methylation (hence the name, short for DNA methyltransferase 3 alpha), which you'll remember is responsible for turning genes on and off, and changes in methylation resulting from its absence then result in more stem cells being produced. Its function is described in detail in Grant A. Challen et al., "Dnmt3a is essential for hematopoietic stem cell differentiation," **Nat. Genet.** 44, 23–31 (2011). DOI: 10.1038/ng.1009 ageless.link/ccese6.

299 **diabetes, and a <u>doubling</u> of the risk of a heart attack or stroke:** Siddhartha Jaiswal et al., "Age-related clonal hematopoiesis associated with adverse outcomes," **N. Engl. J. Med.** 371, 2488–98 (2014). DOI: 10.1056/NEJMoa1408617 ageless.link/ouoyxi.

301 **the mutations present in over 2,500 tumors:** Moritz Gerstung et al., "The evolutionary history

of 2,658 cancers," **Nature** 578, 122–8 (2020). DOI: 10.1038/s41586-019-1907-7 ageless.link/9rgj7s.

302 **A recent proof-of-concept study:** David Fernandez-Antoran et al., "Outcompeting **p53**-mutant cells in the normal esophagus by redox manipulation," **Cell Stem Cell** 25, 329–41 (2019). DOI: 10.1016/j.stem.2019.06.011 ageless .link/xarw3i.

8. REPROGRAMMING AGING

Upgrading our genes

This is an excellent review of the genetics of aging: David Melzer, Luke C. Pilling and Luigi Ferrucci, "The genetics of human aging," **Nat. Rev. Genet.** 21, 88–101 (2019). DOI: 10.1038/s41576-019 -0183-6 ageless.link/t9dut3.

307 **estimate the degree of "heritability" of longevity:** A. M. Herskind et al., "The heritability of human longevity: A population-based study of 2872 Danish twin pairs born 1870–1900," **Hum. Genet.** 97, 319–23 (1996). DOI: 10.1007/ BF02185763 ageless.link/ijjnnc.

307 **A 2018 study, using . . . a genealogy website:** J. Graham Ruby et al., "Estimates of the heritability of human longevity are substantially inflated due to assortative mating," **Genetics** 210, 1109–24 (2018). DOI: 10.1534/genetics.118.301613 ageless .link/p6mjpn.

308 **[centenarians] weigh about the same:**

Swapnil N. Rajpathak et al., "Lifestyle factors of people with exceptional longevity," **J. Am. Geriatr. Soc.** 59, 1509–12 (2011). DOI: 10.1111/j.1532 -5415.2011.03498.x ageless.link/hw9are.

A brief overview of work on genetics and life-styles of centenarians can be found in the first half of this talk: Nir Barzilai, "Can we grow older without growing sicker?" (TEDMED, YouTube, 2017) ageless.link/hza3fp.

308 **[centenarians] spent dramatically less of their lives with disease:** Stacy L. Andersen et al., "Health span approximates life span among many supercentenarians: Compression of morbidity at the approximate limit of life span," **J. Gerontol. A Biol. Sci. Med. Sci.** 67, 395–405 (2012). DOI: 10.1093/gerona/glr223 ageless.link/cmzaqo.

309 **female centenarians outnumber males:** Thomas T. Perls, "Male centenarians: How and why are they different from their female counterparts?," **J. Am. Geriatr. Soc.** 65, 1904–6 (2017). DOI: 10.1111/jgs.14978 ageless.link/a46hmo.

309 **if one of your siblings lives to 100:** Rajpathak et al., 2011 ageless.link/hw9are.

309 **The E4 variant is less common, but it's bad news:** This is a good summary of recent work updating the estimates of risk from APOE variants, particularly the rare **E2** version of the gene: "Rare luck: Two copies of **ApoE2** shield against Alzheimer's," **Alzforum** (2019) ageless.link/yfr6ac.

311 **humans with a favorable FOXO3 variant:**

Cynthia J. Kenyon, "The genetics of aging," **Nature** 464, 504–12 (2010). DOI: 10.1038/nature08980 ageless.link/grpyr3.

312 **a three-year-old girl, part of the Old Order Amish community:** Karen Weintraub, "Gene variant in Amish a clue to better aging," **Genetic Engineering and Biotechnology News** (2018) ageless.link/q3qprd.

314 **people with the mutation lived ten years longer:** Sadiya S. Khan et al., "A null mutation in SERPINE1 protects against biological aging in humans," **Science Advances** 3, eaao1617 (2017). DOI: 10.1126/sciadv.aao1617 ageless.link/qsekck.

315 **A woman in a huge extended family in Colombia:** Sharon Begley, "She was destined to get early Alzheimer's, but didn't," **STAT** (2019) ageless.link/hjynuk.

316 **mice given an extra copy of a gene called Atg5:** Jong-Ok Pyo et al., "Overexpression of Atg5 in mice activates autophagy and extends lifespan," **Nat. Commun.** 4, 2300 (2013) DOI: 10.1038/ncomms3300 ageless.link/cyd9r9.

316 **FGF21 which simulates dietary restriction:** Yuan Zhang et al., "The starvation hormone, fibroblast growth factor-21, extends lifespan in mice," **Elife** 1, e00065 (2012). DOI: 10.7554/eLife.00065 ageless.link/oqp3yy.

316 **One drug now under development:** Joshua Levine et al., "OR22-6 reversal of diet induced metabolic syndrome in mice with an orally active small molecule inhibitor of PAI-1," **J. Endocr. Soc.**

3 (2019). DOI: 10.1210/js.2019-OR22-6 ageless .link/cvbbnm.

319 **AAV gene therapy to treat multiple age-related diseases:** Noah Davidsohn et al., "A single combination gene therapy treats multiple age-related diseases," **Proc. Natl. Acad. Sci. U.S.A.** 47, 23505–11 (2019). DOI: 10.1073/pnas.1910073116 ageless.link/7n97sc.

319 **founded a company in the U.S. called Rejuvenate Bio:** Ryan Cross, "An 'anti-aging' gene therapy trial in dogs begins, and rejuvenate bio hopes humans will be next," **Chemical & Engineering News** (2019) ageless.link/bcbupu.

320 **A study in Dallas, Texas:** Marianne Abifadel et al., "Living the PCSK9 adventure: From the identification of a new gene in familial hyper-cholesterolemia toward a potential new class of anticholesterol drugs," **Curr. Atheroscler. Rep.** 16, 439 (2014). DOI: 10.1007/s11883-014-0439-8 ageless.link/gtc9jy.

320 **a company called Verve Therapeutics:** Ian Sample, "One-off injection may drastically reduce heart attack risk," **Guardian** (10 May 2019) ageless .link/byd76y.

321 **"base editing":** Alexis C. Komor et al., "Programmable editing of a target base in genomic DNA without double-stranded DNA cleavage," **Nature** 533, 420–24 (2016). DOI: 10.1038/ nature17946 ageless.link/xmk79n.

Turning back the epigenetic clock

This article explores the topic of epigenetic reprogramming by profiling one of the scientists at the cutting edge of the technique: Usha Lee McFarling, "The creator of the pig-human chimera keeps proving other scientists wrong," **STAT** (2017) ageless.link/uw74fk.

326 **derived from people as old as 114:** Jieun Lee et al., "Induced pluripotency and spontaneous reversal of cellular aging in supercentenarian donor cells," **Biochem. Biophys. Res. Commun.** 525, 563–569 (2020). DOI: 10.1016/j.bbrc.2020.02.092 ageless .link/rpwt3z.

326 **epigenetic age of zero . . . or a centenarian:** Francesco Ravaioli et al., "Age-related epigenetic derangement upon reprogramming and differentiation of cells from the elderly," **Genes** 9, 39 (2018). DOI: 10.3390/genes9010039 ageless.link/3i4jtt.

326 **the epigenetic reset . . . rejuvenative effects:** Burcu Yener Ilce, Umut Cagin and Acelya Yilmazer, "Cellular reprogramming: A new way to understand aging mechanisms," **Wiley Interdiscip. Rev. Dev. Biol.** 7, e308 (2018). DOI: 10.1002/ wdev.308 ageless.link/6ewuqx.

328 **Dolly started life with a biological handicap:** Kevin Sinclair, "Dolly's 'sisters' show cloned animals don't grow old before their time," **The Conversation** (2016) ageless.link/xdyba3.

José Cibelli, "More lessons from Dolly the

sheep: Is a clone really born at age zero?," **The Conversation** (2017) ageless.link/hgwufq.

329 **eventually made it to the sixth generation:** Sayaka Wakayama et al., "Successful serial re-cloning in the mouse over multiple generations," **Cell Stem Cell** 12, 293–7 (2013). DOI: 10.1016/j.stem.2013.01.005 ageless.link/kxyfii.

330 **The jellyfish Turritopsis dohrnii:** Nathaniel Rich, "Can a jellyfish unlock the secret of immortality?," **New York Times** (28 November 2012) ageless.link/7zcdy4.

331 **a more subtle approach enjoyed greater success:** Alejandro Ocampo et al., "In vivo ameliora-tion of age-associated hallmarks by partial re-programming," **Cell** 167, 1719–733.e12 (2016). DOI: 10.1016/j.cell.2016.11.052 ageless.link/cssud4.

332 **transient activation of the Yamanaka factors and:** Tapash Jay Sarkar et al., "Transient non-integrative expression of nuclear reprogramming factors promotes multifaceted amelioration of aging in human cells," **Nat. Commun.** 11, 1545 (2020). DOI: 10.1038/s41467-020-15174-3 ageless.link/96ac3p.

333 **reprogramming . . . eye injuries in middle-aged mice:** Yuancheng Lu et al., "Reversal of aging- and injury-induced vision loss by Tet-dependent epigenetic reprogramming," **bioRxiv** (2019). DOI: 10.1101/710210 ageless.link/7zv3rh.

333 **[inducing pluripotency is] a multistep process:** Nelly Olova et al., "Partial reprogramming induces

a steady decline in epigenetic age before loss of somatic identity," **Aging Cell** 18, e12877 (2019). DOI: 10.1111/acel.12877 ageless.link/yo3wwk.

334 **a different cocktail of genes turns . . . into another:** Deepak Srivastava and Natalie DeWitt, "In vivo cellular reprogramming: the next generation," **Cell** 166, 1386–96 (2016). DOI: 10.1016/j.cell.2016.08.055 ageless.link/xor74i.

335 **"chemically induced reprogramming":** Dhruba Biswas and Peng Jiang, "Chemically induced reprogramming of somatic cells to pluripotent stem cells and neural cells," **Int. J. Mol. Sci.** 17, 226 (2016). DOI: 10.3390/ijms17020226 ageless.link/7nhpma.

336 **One suggested . . . embryonic–fetal transition:** Michael D. West et al., "Use of deep neural network ensembles to identify embryonic–fetal transition markers: repression of COX7A1 in embryonic and cancer cells," **Oncotarget** 9, 7796–811 (2018). DOI: 10.18632/oncotarget.23748 ageless.link/zc6zye.

Reprogramming biology and curing aging

A brief introduction to the idea of using systems biology for medicine can be found in Rolf Apweiler et al., "Whither systems medicine?," **Exp. Mol. Med.** 50, e453 (2018). DOI: 10.1038/emm.2017.290 ageless.link/vfusyd.

345 **simulate a bacterium called Mycoplasma genitalium:** Jonathan R. Karr et al., "A whole-cell computational model predicts phenotype

from genotype," **Cell** 150, 389–401 (2012). DOI: 10.1016/j.cell.2012.05.044 ageless.link/ cecsmo.

345 **stop the virus from rapidly evolving resistance:** A. S. Perelson et al., "HIV-1 dynamics in vivo: Virion clearance rate, infected cell life-span, and viral generation time," **Science** 271, 1582–6 (1996). DOI: 10.1126/science.271.5255.1582 ageless.link/ub43sm.

346 **train a computer model to recognize . . . DR mimetics:** Diogo G. Barardo et al., "Machine learning for predicting lifespan-extending chemical compounds," **Aging** 9, 1721–37 (2017). DOI: 10.18632/aging.101264 ageless.link/z67qqd.

346 **in 2019, a whole-genome . . . $1,000:** The cost of sequencing a human genome (National Human Genome Research Institute, 2019) ageless.link/ 79qfqn.

347 **Computing power has doubled every two years:** Max Roser and Hannah Ritchie, "Technological progress," **Our World in Data** (2013) ageless.link /capdvn.

9. THE QUEST FOR A CURE

357 **discovered in 2013 in <u>C. elegans</u>:** Di Chen et al., "Germline signaling mediates the synergistically prolonged longevity produced by double mutations in **daf-2** and **rsks-1** in **C. elegans**," **Cell Rep.** 5, 1600–1610 (2013). DOI: 10.1016/ j.celrep.2013.11.018 ageless.link/qhwo37.

10. HOW TO LIVE LONG ENOUGH
TO LIVE EVEN LONGER

363 **One study looking at 100,000 health professionals in the U.S.:** Yanping Li et al., "Healthy lifestyle and life expectancy free of cancer, cardiovascular disease, and type 2 diabetes: Prospective cohort study," **BMJ** 368, l6669 (2020). DOI: 10.1136/bmj.l6669 ageless.link/3i3g3w.

364 **40 percent of cancer:** Statistics on preventable cancers (Cancer Research UK, 2015) ageless.link/jtbsb9.

364 **a staggering 80 percent of cardiovascular disease:** Cardiovascular disease data and statistics (World Health Organization, 2020) ageless.link/p3tz36.

365 **those who are . . . who benefit:** Gaëlle Deley et al., "Physical and psychological effectiveness of cardiac rehabilitation: Age is not a limiting factor!," **Can. J. Cardiol.** 35, 1353–8 (2019). DOI: 10.1016/j.cjca.2019.05.038 ageless.link/r6dzqn.

1. Don't smoke

365 **Smokers can't . . . die young:** Jha, 2009 ageless.link/fjnhnq.

366 **They leave a specific "mutational signature":** Yoshida et al., 2020 ageless.link/7yisot.

367 **Inflammation falls rapidly after giving up:** Virginia Reichert et al., "A pilot study to examine the effects of smoking cessation on serum

markers of inflammation in women at risk for cardiovascular disease," **Chest** 136, 212–19 (2009). DOI: 10.1378/chest.08-2288 ageless.link/hdjg9s.

2. Don't eat too much

367 **Getting a balanced diet . . . lifespan:** Lukas Schwingshackl et al., "Food groups and risk of all-cause mortality: A systematic review and meta-analysis of prospective studies," **Am. J. Clin. Nutr.** 05, 462–73 (2017). DOI: 10.3945/ajcn.117.153148 ageless.link/4bfurj.

368 **The observational . . . of vegetarianism:** Monica Dinu et al., "Vegetarian, vegan diets and multiple health outcomes: A systematic review with meta-analysis of observational studies," **Crit. Rev. Food Sci. Nutr.** 57, 3640–49 (2017). DOI: 10.1080/10408398.2016.1138447 ageless.link/6htpi3.

368 **fruit and vegetables . . . diversity of your microbiome:** Society for Applied Microbiology, 2019 ageless.link/enkq6q.

369 **our bodies overcompensate . . . making us healthier:** Tae Gen Son, Simonetta Camandola and Mark P. Mattson, "Hormetic dietary phytochemicals," **Neuromolecular Med.** 10, 236–46 (2008). DOI: 10.1007/s12017-008-8037-y ageless.link/6u6wox.

369 **excess fat is bad for you:** Dagfinn Aune et al., "BMI and all cause mortality: Systematic review and non-linear dose-response meta-analysis of 230 cohort studies with 3.74 million deaths among

30.3 million participants," **BMJ** 353, i2 56 (2016). DOI: 10.1136/bmj.i2156 ageless.link/b4nzgu.

370 **being heavier still can knock a full decade off:** The discussion section of this paper summarizes a number of different studies on the effect of weight on lifespan: Steven A. Grover et al., "Years of life lost and healthy life-years lost from diabetes and cardiovascular disease in overweight and obese people: A modelling study," **Lancet Diabetes Endocrinol** 3, 114–22 (2015). DOI: 10.1016/S2213-8587(14)70229-3 ageless.link/dsg3py.

370 **obese people . . . incur greater health expenses:** Eric A. Finkelstein et al., "The lifetime medical cost burden of overweight and obesity: Implications for obesity prevention," **Obesity** 6, 843–8 (2008). DOI: 10.1038/oby.2008.290 ageless.link/9aqtvu.

370 **Visceral fat seems to be by far the worse:** For a readable summary of the science of inflammation and fat, see "Taking aim at belly fat" (Harvard Health, 2010) ageless.link/e6do9f.

A more technical overview can be found in Volatiana Rakotoarivelo et al., "Inflammatory cytokine profiles in visceral and subcutaneous adipose tissues of obese patients undergoing bariatric surgery reveal lack of correlation with obesity or diabetes," **EBioMedicine** 30, 237–47 (2018). DOI: 10.1016/j.ebiom.2018.03.004 ageless.link/67vyza.

372 **Waist-to-height ratio improves on [BMI]:** Márcia Mara Corrêa et al., "Performance of the waist-to-height ratio in identifying obesity and

predicting non-communicable diseases in the elderly population: A systematic literature review," **Arch. Gerontol. Geriatr.** 65, 74–82 (2016). DOI: 10.1016/j.archger.2016.03.021 ageless.link/kn7b97.

372 **losing weight . . . get these problems under control:** "Does weight loss cure type 2 diabetes?" (British Heart Foundation, 2017) ageless.link/94ty9p.

372 **cutting back . . . health benefits:** Manuela Aragno and Raffaella Mastrocola, "Dietary sugars and endogenous formation of advanced glycation endproducts: Emerging mechanisms of disease," **Nutrients** 9 (2017). DOI: 10.3390/nu9040385 ageless.link/xbx6zn.

373 **dietary AGEs . . . to avoid:** Jaime Uribarri et al., "Advanced glycation end products in foods and a practical guide to their reduction in the diet," **J. Am. Diet. Assoc.** 0, 9—6.e 2 (2010). DOI: 10.1016/j.jada.2010.03.018 ageless.link/qxtoer. Extance, 2018 ageless.link/ep3o7t.

373 **with a total of just under 200 monkeys:** An accessible write-up of the studies which goes into detail about the differences between them can be found here: Gifford, 2012 ageless.link/kcc4qs.

374 **the NIA monkeys . . . had statistically indistinguishable lifespans:** Mattison et al., 2017 ageless.link/jnaqjv.

375 **Hawkish proponents of DR contend:** This is one example—a very detailed article making the case that DR does work using evidence from monkeys

and other primates by Michael Rae, a board member of the CR Society International: Michael Rae, "CR in nonhuman primates: A muddle for monkeys, men, and mimetics" (SENS Research Foundation, 2013) ageless.link/794i74.

375 **There have been studies in humans:** Kraus et al., 2019 ageless.link/t6tm4m.

376 **the unusual longevity of people in Okinawa:** Natalia S. Gavrilova and Leonid A. Gavrilov, "Comments on dietary restriction, Okinawa diet and longevity," **Gerontology** 58, 221–23 (2012). DOI: 10.1159/000329894 ageless.link/jkkwhw.

376 **DR mice with flu die more often:** Elizabeth M. Gardner, "Caloric restriction decreases survival of aged mice in response to primary influenza infection," **J. Gerontol. A Biol. Sci. Med. Sci.** 60, 688–94 (2005). DOI: 10.1093/gerona/60.6.688 ageless.link/vw6q4r.

376 **a few participants . . . stop due to anemia:** Eric Ravussin et al., "A 2-year randomized controlled trial of human caloric restriction: Feasibility and effects on predictors of health span and longevity," **J. Gerontol. A Biol. Sci. Med. Sci.** 70, 097–04 (2015). DOI: 10.1093/gerona/glv057 ageless.link /ci3m6v.

377 **When you eat as well as what you eat:** This is a provocative review advocating the benefits of intermittent fasting, and suggesting how to go about it if you want to try: Rafael de Cabo and Mark P. Mattson, "Effects of intermittent fasting on health, aging, and disease," **N. Engl. J. Med.**

38, 254–5 (2019). DOI: 10.1056/NEJMra1905136 ageless.link/3pgwep.

3. Get some exercise

379 **Exercise is good for your health:** To illustrate how sure we are that exercise is good for you, this paper isn't just a systematic review—where researchers try to pull together every study that's ever been done on a topic and assess them all together—but a systematic review **of systematic reviews** that shows this: Darren E. R. Warburton and Shannon S. D. Bredin, "Health benefits of physical activity: A systematic review of current systematic reviews," **Curr. Opin. Cardiol.** 32, 541–56 (2017). DOI: 10.1097/HCO.0000000000000437 ageless .link/9mef3o.

379 **If your lifestyle is entirely sedentary:** Erika Rees-Punia et al., "Mortality risk reductions for replacing sedentary time with physical activities," **Am. J. Prev. Med.** 56, 736–41 (2019). DOI: 10.1016/ j.amepre.2018.12.006 ageless.link/xrfogk.

380 **Benefits beyond this are unclear:** Ulf Ekelund et al., "Dose-response associations between accelerometry measured physical activity and sedentary time and all cause mortality: Systematic review and harmonised meta-analysis," **BMJ** 366, l4570 (2019). DOI: 10.1136/bmj.l4570 ageless.link/ 7khsm6.

380 **Olympic athletes have lower mortality:** Taro Takeuchi et al., "Mortality of Japanese Olympic athletes: 1952–2017 cohort study," **BMJ Open**

Sport Exerc. Med. 5, e000653 (2019). DOI: 10.1136/bmjsem-2019-000653 ageless.link/qkghkf.

380 **chess champions also live longer:** An Tran-Duy, David C. Smerdon and Philip M. Clarke, "Longevity of outstanding sporting achievers: Mind versus muscle," **PLoS One** 13, e0196938 (2018). DOI: 10.1371/journal.pone.0196938 ageless.link/xsw9i7.

380 **and Nobel Prize winners:** Matthew D. Rablen and Andrew J. Oswald, "Mortality and immortality: The Nobel Prize as an experiment into the effect of status upon longevity," **J. Health Econ.** 27, 1462–71 (2008). DOI: 10.1016/j.jhealeco.2008.06.001 ageless.link/fbjyns.

381 **We lose about 5 percent of our muscle mass and 10 percent of our strength every decade:** W. Kyle Mitchell et al., "Sarcopenia, dynapenia, and the impact of advancing age on human skeletal muscle size and strength; a quantitative review," **Front. Physiol.** 3, 260 (2012). DOI: 10.3389/fphys.2012.00260 ageless.link/agabb4.

381 **exercise programs . . . 90-somethings:** Eduardo L. Cadore et al., "Multicomponent exercises including muscle power training enhance muscle mass, power output, and functional outcomes in institutionalized frail nonagenarians," **Age** 36, 773–85 (2014). DOI: 10.1007/s11357-013-9586-z ageless.link/3bcah673–85 (20 4). DOI: 10.1007/s11357-013-9586-z ageless.link/3bcah6.

4. Get seven to eight hours of sleep a night

382 **seven or eight hours . . . optimal for health:** Xiaoli Shen, Yili Wu and Dongfeng Zhang, "Nighttime sleep duration, 24-hour sleep duration and risk of all-cause mortality among adults: A meta-analysis of prospective cohort studies," **Sci. Rep.** 6, 21480 (2016). DOI: 10.1038/srep21480 ageless.link/mnz6j3.

383 **while we sleep, our brains . . . spring-clean:** Ehsan Shokri-Kojori et al., "β-amyloid accumulation in the human brain after one night of sleep deprivation," **Proc. Natl. Acad. Sci. U.S.A.** 115, 4483–8 (2018). DOI: 10.1073/pnas.1721694115 ageless.link/ixiidn.

383 **the clouding and discoloration of the lens:** Line Kessel et al., "Sleep disturbances are related to decreased transmission of blue light to the retina caused by lens yellowing," **Sleep** 34, 1215–19 (2011). DOI: 10.5665/SLEEP.1242 ageless.link/eaykuc.

5. Get vaccinated and wash your hands

385 **deaths from heart attacks:** Alejandra Pera et al., "Immunosenescence: Implications for response to infection and vaccination in older people," **Maturitas** 82, 50–55 (2015). DOI: 10.1016/j.maturitas.2015.05.004 ageless.link/jg7nsn.

386 **additional, indirect effects on life expectancy:** Caleb E. Finch and Eileen M. Crimmins, "Inflammatory exposure and historical changes in human

life-spans," **Science** 305, 1736–9 (2004). DOI: 10.1126/science.1092556 ageless.link/uiaa3d.

6. Take care of your teeth

388 **people who brushed their teeth . . . were at lower risk:** Cesar de Oliveira, Richard Watt and Mark Hamer, "Toothbrushing, inflammation, and risk of cardiovascular disease: Results from Scottish Health Survey," **BMJ** 340, c2451 (2010). DOI: 10.1136/bmj.c2451 ageless.link/4igja4.

388 **types of bacteria . . . an effect:** Chung-Jung Chiu, Min-Lee Chang and Allen Taylor, "Associations between periodontal microbiota and death rates," **Sci. Rep.** 6, 35428 (2016). DOI: 10.1038/srep35428 ageless.link/st9goi.

7. Wear sunscreen

389 **Getting sunburn just once every two years:** Leslie K. Dennis et al., "Sunburns and risk of cutaneous melanoma: Does age matter? A comprehensive meta-analysis," **Ann. Epidemiol.** 18, 614–27 (2008). DOI: 10.1016/j.annepidem.2008.04.006 ageless.link/yd4jxa.

8. Monitor your heart rate and blood pressure

392 **Globally, around 40 percent of people over the age of 25 suffer from it:** "Raised blood pressure" (World Health Organization Global Health Observatory, 2015) ageless.link/bzteab.

392 **every additional 20/10 roughly doubles the risk:** Sarah Lewington et al., "Age-specific relevance

of usual blood pressure to vascular mortality: A meta-analysis of individual data for one million adults in 61 prospective studies," **Lancet** 360, 1903–13 (2002). DOI: 10.1016/s0140 -6736(02)11911-8 ageless.link/tknbz6.

392 **If you consistently get results of 140/90 or more:** "High blood pressure (hypertension)" (NHS, 2019) ageless.link/jy364p.

"New ACC/AHA high blood pressure guidelines lower definition of hypertension" (American College of Cardiology, 2017) ageless.link/ mtpxoi.

393 **Intriguingly . . . not just heart disease:** Aune et al., 2017 ageless.link/9hukvg.

9. Don't bother with supplements

393 **half of U.S. adults . . . regularly:** Elizabeth D. Kantor et al., "Trends in dietary supplement use among U.S. adults from 1999–2012," **JAMA** 316, 1464–74 (2016). DOI: 10.1001/jama.2016.14403 ageless.link/sbmuq9.

10. Don't bother with longevity drugs—yet

394 **aspirin comes with a risk . . . bleeding:** Donna K. Arnett et al., "2019 ACC/AHA guideline on the primary prevention of cardiovascular disease: A report of the American College of Cardiology/American Heart Association task force on clinical practice guidelines," **J. Am. Coll. Cardiol.** 74, e177–e232 (2019). DOI: 10.1016/ j.jacc.2019.03.010 ageless.link/ttziau.

395 **humans can get . . . cholesterol:** Charles Faselis et al., "Is very low LDL-C harmful?," **Curr. Pharm. Des.** 24, 3658–64 (2018). DOI: 10.2174/1381612 824666181008110643 ageless.link/7uqaqe.

11. Be a woman

395 **being a born woman . . . expectancy:** This was calculated using WHO GBD statistics. You can read about the calculations at ageless.link/tv7grc.

395 **However, there are also a few biological explanations:** Steven N. Austad and Kathleen E. Fischer, "Sex differences in lifespan," **Cell Metab.** 23, 1022–33 (2016). DOI: 10.1016/ j.cmet.2016.05.019 ageless.link/xonwam.

396 **whichever sex . . . chromosomes:** Zoe A. Xirocostas, Susan E. Everingham and Angela T. Moles, "The sex with the reduced sex chromosome dies earlier: A comparison across the tree of life," **Biol. Lett.** 16, 20190867 (2020). DOI: 10.1098/ rsbl.2019.0867 ageless.link/vvqsmi.

396 **mitochondria may play a role:** M. Florencia Camus, David J. Clancy and Damian K. Dowling, "Mitochondria, maternal inheritance, and male aging," **Curr. Biol.** 22, 1717–21 (2012). DOI: 10.1016/j.cub.2012.07.018 ageless.link/ jedc3a.

397 **worse health on average:** Susan C. Alberts et al., **The Male-Female Health-Survival Paradox: A Comparative Perspective on Sex Differences in Aging and Mortality** (National Academies Press [U.S.], 2014) ageless.link/gkjfgw.

398 **There is still some debate:** This is one example of data that go against the hypothesis. As we discussed in Chapter 1, estimating healthy life expectancy is far more complicated than life expectancy overall.

Healthy life expectancy (HALE): Data by country (World Health Organization Global Health Observatory, 2018) ageless.link/mbznxr.

398 **37 percent of the centenarian men didn't suffer:** Nisha C. Hazra et al., "Differences in health at age 100 according to sex: Population-based cohort study of centenarians using electronic health records," **J. Am. Geriatr. Soc.** 63, 1331–7 (2015). DOI: 10.1111/jgs.13484 ageless.link/bkzvue.

11. FROM SCIENCE TO MEDICINE

This review examines some of the scientific and policy shifts necessary to realize the potential of biogerontology. It also introduces yet another classification of age-related changes, the "pillars of aging"!

Brian K. Kennedy et al., "Geroscience: Linking aging to chronic disease," **Cell** 159, 709–13 (2014). DOI: 10.1016/j.cell.2014.10.039 ageless.link/hnoqys.

401 **90 percent . . . had heard only a little:** "Living to 120 and beyond: Americans' views on aging, medical advances and radical life extension" (Pew Research Center, 2013) ageless.link/jrmgc3.

402 **there is little incentive to deviate:** This paper tracks changes in scientists' research interest over

time by monitoring their publications, and finds that substantial changes in field are rare: Tao Jia, Dashun Wang and Boleslaw K. Szymanski, "Quantifying patterns of research-interest evolution," **Nature Human Behavior** 1, 0078 (2017). DOI: 10.1038/s41562-017-0078 ageless.link/yo7zw3.

402 **The National Institute on Aging . . . budget of $2.6 billion:** The figures in this and subsequent paragraphs have been compiled from the NIA, NIH, NCI and CMS, and can be found at ageless .link/7679wa.

403 **It's a biogerontology in-joke:** This delightfully curmudgeonly comment by Leonard Hayflick (of the Hayflick limit, biogerontology pioneer and founding member of the Council of the NIA) is a great example of the infuriation this causes: Leonard Hayflick, "Comment on 'We have a budget for FY 2019!'" (2018) ageless.link/9p6cw3.

404 **worth seven trillion dollars over 50 years:** Dana P. Goldman et al., "Substantial health and economic returns from delayed aging may warrant a new focus for medical research," **Health Aff.** 32, 1698–1705 (2013). DOI: 10.1377/hlthaff.2013.0052 ageless.link/ctacos.

406 **a comparison . . . metformin:** This is the original study comparing metformin to sulphonylureas and healthy people taking neither: C. A. Bannister et al., "Can people with type 2 diabetes live longer than those without? A comparison of mortality in people initiated with metformin or sulphonylurea

monotherapy and matched, non-diabetic controls," **Diabetes Obes. Metab.** 16, 1165–73 (2014). DOI: 10.1111/dom.12354 ageless.link/oxih3v.

This paper by Nir Barzilai rounds up the evidence for metformin and making the case for its anti-aging properties: Nir Barzilai et al., "Metformin as a tool to target aging," **Cell Metab.** 23, 1060–65 (2016). DOI: 10.1016/j.cmet.2016.05.011 ageless.link/yv7ssx.

407 **That's the aim of the TAME trial:** Barzilai discusses TAME in the last part of this talk: Barzilai, 2017 ageless.link/awkcqw.

410 **quickly add up a patient's epigenetic age:** Steve Horvath and Kenneth Raj, "DNA methylation-based biomarkers and the epigenetic clock theory of aging," **Nat. Rev. Genet.** 19. 371–84 (2018). DOI: 10.1038/s41576-018-0004-3 ageless.link/jyhwdv.

410 **A new version . . . clock:** Ake T. Lu et al., "DNA methylation GrimAge strongly predicts lifespan and healthspan," **Aging** 11, 303–27 (2019). DOI: 10.18632/aging.101684 ageless.link/ijx34n.

411 **guess people's ages . . . photographs:** Kaare Christensen et al., "Perceived age as clinically useful biomarker of aging: Cohort study," **BMJ** 339, b5262 (2009). DOI: 10.1136/bmj.b5262 ageless.link/c7bbfy.

411 **The next step is to automate this:** Weiyang Chen et al., "Three-dimensional human facial morphologies as robust aging markers," **Cell Res.** 25,

574–87 (2015). DOI: 10.1038/cr.2015.36 ageless
.link/4h3ivk.

411 **A team is . . . for mice:** Alex Zhavoronkov and
Polina Mamoshina, "Deep aging clocks: The
emergence of AI-based biomarkers of aging and
longevity," **Trends Pharmacol. Sci.** 40, 546–9
(2019). DOI: 10.1016/j.tips.2019.05.004 ageless
.link/uvip6c.

412 **epigenetic clocks . . . slowed by dietary
restriction:** Tina Wang et al., "Epigenetic aging
signatures in mice livers are slowed by dwarfism,
calorie restriction and rapamycin treatment,"
Genome Biol. 18, 57 (2017). DOI: 10.1186/
s13059-017-1186-2 ageless.link/9sgahr.

413 **A similar result in rhesus monkeys:** Shinji
Maegawa et al., "Caloric restriction delays age-related methylation drift," **Nat. Commun.** 8, 539
(2017). DOI: 10.1038/s41467-017-00607-3
ageless.link/migjww.

413 **could theoretically be completed:** Josh
Mitteldorf, "The mother of all clinical trials," part
1 (2018) ageless.link/s9p3fs.

414 **New treatments . . . rarely trialed in the elderly:**
Antonio Cherubini et al., "Fighting against age
discrimination in clinical trials," **J. Am. Geriatr.
Soc.** 58, 1791–6 (2010). DOI: 10.1111/j.1532
-5415.2010.03032.x ageless.link/io4zwa.

415 **The same problem . . . mouse studies:** Kennedy
et al., 2014 ageless.link/hnoqys.

416 **flu shots . . . the morning:** Joanna E. Long

et al., "Morning vaccination enhances antibody response over afternoon vaccination: A cluster-randomized trial," **Vaccine** 34, 2679–85 (2016). DOI: 10.1016/j.vaccine.2016.04.032 ageless.link/77mqxq.

417 **tougher to raise funding for:** According to one example from a review of senescent cells in aging, the lack of studies "is not because those in the field do not recognize the utility of such work; rather, it is that studies of this type are considered—in the words of one anonymous but unsympathetic referee—'boring descriptive work.'"

Richard G. A. Faragher et al., "Senescence in the aging process," **F1000Res.** 6, 1219 (2017). DOI: 10.12688/f1000research.10903.1 ageless.link/q6yvhy.

420 **a biotech CEO going to a Colombian clinic:** Nicola Davis and Dara Mohammadi, "Can this woman cure aging with gene therapy?," **Guardian** (24 July 2016) ageless.link/m4u9yb.

Index

ABOUT THE AUTHOR

After obtaining a PhD in physics from the University of Oxford, ANDREW STEELE decided that aging was the most important scientific challenge of our time and switched fields to computational biology. He worked at the Francis Crick Institute, using machine learning to decode our DNA and predict heart attacks using patients' medical records. He is now a full-time science writer and presenter based in London. He has appeared on the BBC and on Discovery, Inc.'s Science Channel.

andrewsteele.co.uk
Twitter: @statto
Instagram: @andrewjsteele